Entomology in Focus

Volume 6

Series editor
Simon L. Elliot, Viçosa, Minas Gerais, Brazil

Insects are fundamentally important in the ecology of terrestrial habitats. What is more, they affect diverse human activities, notably agriculture, as well as human health and wellbeing. Meanwhile, much of modern biology has been developed using insects as subjects of study.

To reflect this, our aim with Entomology in Focus is to offer a range of titles that either capture different aspects of the diverse biology of insects or their management, or that offer updates and reviews of particular species or taxonomic groups that are important for agriculture, the environment or public health.

The series results from an agreement between Springer and the Entomological Society of Brazil (SEB) and as such may lean towards tropical entomology. The aim throughout is to provide reference texts that are simple in their conception and organization but that offer up-to-date syntheses of the respective areas, offer suggestions of future directions for research (and for management where relevant) and that do not shy away from offering considered opinions.

More information about this series at http://www.springer.com/series/10465

Antônio Ricardo Panizzi • Tiago Lucini
Paula Levin Mitchell

Electronic Monitoring of Feeding Behavior of Phytophagous True Bugs (Heteroptera)

Springer

Antônio Ricardo Panizzi
Embrapa Trigo
Passo Fundo, Rio Grande do Sul, Brazil

Tiago Lucini
Embrapa Trigo
Passo Fundo, Rio Grande do Sul, Brazil

Paula Levin Mitchell
Department of Biology
Winthrop University
Rock Hill, SC, USA

ISSN 2405-853X ISSN 2405-8548 (electronic)
Entomology in Focus
ISBN 978-3-030-64676-9 ISBN 978-3-030-64674-5 (eBook)
https://doi.org/10.1007/978-3-030-64674-5

This Springer imprint is published by the registered company Springer Nature Switzerland AG
The registered company address is: Gewerbestrasse 11, 6330 Cham, Switzerland

Foreword

In the late 1950s, Donald L. McLean, a young, new faculty member at the University of California (UC) stationed at the university's research center in Salinas, California, accepted a challenge from one of his former Ph.D. committee members at UC Berkeley, the well-known vector entomologist, Ned Sylvester. The challenge was to develop some kind of electronic instrument that could measure within a second the start and end of an aphid's stylet penetration. Such behavior could not be directly visualized because the aphid's stylets were hidden within its labium, and brief proboscis contact with a plant without stylet insertion was common for aphids. Sylvester and McLean were both studying acquisition and inoculation of nonpersistent plant viruses by aphid vectors, a process that could be completed within a few seconds. They needed a better way to time that process.

McLean had an interest in electronics and was a ham radio enthusiast. He accepted the challenge and, with the help of his newly hired technician, Marv Kinsey, designed such a device as well as the many insect-manipulating, tethering, and research protocols to go with it. Many of these research protocols are still used today. After transferring to the then-new "farm campus," later named UC Davis, McLean and Kinsey continued their iterative design efforts. They quickly understood the important implications for aphid biology when they saw that, not only were the beginning and ending of stylet penetration portrayed in their electronic output, but also highly meaningful waveforms representing stylet penetration behaviors, in between. McLean and Kinsey published their first study of "electronic measurement" of aphid feeding in 1964 in *Nature* (Backus 1994; McLean and Kinsey 1964). Their device was similar to an AM radio, with AC applied signal and low amplifier sensitivity. In the nearly 60 years since their insightful and empirically successful, landmark publication, over 670 papers have now been published using remarkably similar (yet also more advanced) electronics and methods (Backus et al. 2021). In many ways, their idea has revolutionized the study of hemipteran feeding. It has certainly made possible nearly all of our present understanding of the role of vector feeding in vector-mediated transmission of plant pathogens (Backus et al. 2019; Brown 2016).

Fourteen years after their first publication, in September 1978, I arrived in the UC Davis lab of McLean and Kinsey as a young graduate student. Only a few months before, another, somewhat older graduate student on the other side of the planet, at Wageningen University in the Netherlands, had just published his own paper that would modernize and ultimately revolutionize the work of McLean and Kinsey. Fred (Freddy) W. Tjallingii had trained in insect electrophysiology under the renowned Wageningen scientist Louis M. Schoonhoven. Tjallingii had a very different perspective and training in electronics, but had read the earlier works of McLean and Kinsey and also became fascinated with the idea of electronically recording aphid feeding. Tjallingii redesigned key aspects of the instrument (notably higher amplifier sensitivity and DC applied signal). He also renamed it the electrical penetration graph, or EPG. Over the next several years, Tjallingii's insightful findings would come to form the theoretical underpinnings of the future science of EPG, especially his concepts of the R and emf components, the "electrical origins" of waveforms.

Although my Ph.D. dissertation research centered on ultrastructural anatomy of sensory organs in leafhoppers, I was intrigued by the old monitoring equipment sitting in the corner of the McLean and Kinsey lab. Five years after I started my graduate program, I finally had a chance to work with an electronic monitor during a short post-doc in the McLean and Kinsey lab; I fell head-over-heels in love with its waveforms. A year later, in 1984, when I was hired as a new assistant professor of entomology at the University of Missouri-Columbia (UMC), I brought a monitor with me. Ever since, EPG has been at the heart of all of my research.

As a graduate student, I loved insect behavior and neurophysiology, as well as insect anatomy. I took a class in electrophysiology, so when I later read Tjallingii's electrophysiological methods (Tjallingii 1978, 1985), I accepted and understood his early findings on the then-revolutionary aphid potential drop waveform. However, because I was studying leafhoppers and other insects larger than aphids, I also saw the value of the original AC instrument. In 1987, my first graduate student, Astri Wayadande (who later joined the faculty at Oklahoma State University), and I had the opportunity to travel to Wageningen to attend Tjallingii's first-ever EPG workshop. I learned there about his ideas on emf responsiveness curves. That's when I realized how we could make EPG more flexible, to tailor the best settings for recording big insects like heteropterans, and in fact, any kind of arthropod (Backus et al. 2019). From that moment forward, I have been driven to bring unity to the science of EPG through the development of a third type of monitor, a "universal" instrument. I also wanted to honor the contributions of all of its founders, McLean, Kinsey, and Tjallingii.

It took 22 years of iterative design and evaluation research, working with several electrical engineering and electronics collaborators, to finally develop (Bill Bennett and Mike Devaney of UMC), publish (Backus and Bennett 2009; Backus et al.

2000), patent (Backus and Bennett 2011), and commercialize (Andy Dowell of EPG Technologies, Inc., Gainesville, FL) the AC-DC electropenetrograph (or AC-DC EPG monitor). This work continued after I left UMC and joined the USDA Agricultural Research Service. We also took the opportunity to modernize the name of the science to electropenetrography (also abbreviated EPG). Along the way, I helped organize, edit, and publish the proceedings of two symposia, one on AC EPG in honor of McLean and Kinsey (Ellsbury et al. 1994), and another on a holistic summary of the advances of both AC and DC EPG, and controversies of the time (Walker and Backus 2000). Until now, these were the only books published in EPG science. It was thus with tremendous excitement and pleasure that I accepted the authors' invitation to write this foreword for the first, hardcover, scientific book centered on EPG research.

As more fully explained in Chap. 1, even today, the majority of species studied with EPG remain tiny insects – aphids and other members of the hemipteran suborder Sternorrhyncha. Yet, more and more scientists are recognizing that the tremendous advancements in aphidology using EPG can and should be replicated with other groups of arthropods. Heteropterists are fortunate that true bug waveforms, like those of Sternorrhyncha, are remarkably similar to one another; this makes their interpretation easier than, for example, those of Auchenorrhyncha. After 45 years of study, there is still no overarching synthesis for auchenorrhynchan EPG. Consequently, this book will be a very useful resource for new EPG researchers tackling yet-unstudied species of Heteroptera. The comprehensive reviews of heteropteran mouthparts, feeding strategies, and feeding physiology in Chaps. 2 and 3 put all relevant, related research in one place, setting the stage for Chaps. 4 and 5 on electronic monitoring of true bugs (EPG plus use of other devices). The encyclopedic analyses of studies to date in the latter two chapters organize large amounts of detailed information from very disparate studies in the primary literature. These summaries thus put all the pertinent research clearly within easy grasp of novice EPG'ers. The unique comparison of EPG waveforms among heteropteran families at the end of Chap. 5 synthesizes all of the analyzed information, in the context of feeding strategies and plant damage caused. Chapter 6 then summarizes the many advancements in techniques and protocols developed to make possible EPG studies with such large insects (compared with all other insects EPG-recorded). Together, Chaps. 2, 3, 4, 5, and 6 present a tour-de-force to launch the discussion in Chaps. 7 and 8 about how further heteropteran EPG research could dramatically improve management of Heteroptera, among the most important group of emerging pests in the new world of climate change. Ultimately, I am confident that this book will both stimulate and accelerate EPG research and positively impact crop protection and global agriculture.

May all readers of this book be inspired to begin their own life-long infatuation with EPG waveforms.

Elaine A. Backus
Parlier, CA, USA

References

Backus EA (1994) History, development, and applications of the AC electronic monitoring system for insect feeding, In: Ellsbury MM, Backus EA, Ullman DL (eds) History, development, and application of AC electronic insect feeding monitors. Entomol Soc Am, Lanham, MD, pp 1–51

Backus EA, Bennett WH (2009) The AC-DC correlation monitor: new EPG design with flexible input resistors to detect both R and emf components for any piercing-sucking hemipteran. J Insect Physiol 55:869–884

Backus EA, Bennett WH (2011) Electrical penetration graph system, patent no. 8,004,292 Federal Register (ed. by U.S. Patent Office) assignee: USDA, U.S.A

Backus EA, Cervantes FA, Guedes RNC, Li AY, Wayadande AC (2019) AC–DC electropenetrography for in-depth studies of feeding and oviposition behaviors. Ann Entomol Soc Am 112:236–248

Backus EA, Devaney MJ, Bennett WH (2000) Comparison of signal processing circuits among seven AC electronic monitoring systems for their effects on the emf and R components of aphid (Homoptera: Aphididae) waveforms. In: Walker GP, Backus EA (eds) Principles and applications of electronic monitoring and other techniques in the study of homopteran feeding behavior. Entomol Soc Am, Lanham, MD, pp 102–143

Backus EA, Guedes RNC, Reif KE (2021) AC-DC electropenetrography: fundamentals, controversies, and perspectives for arthropod pest management. Pest Manage Sci 77:1132–1149

Brown JK (ed) (2016) Vector-mediated transmission of plant pathogens. The Phytopath Soc Am, St. Paul, MN

Ellsbury MM, Backus EA, Ullman DE (eds) (1994) History, development and application of AC electronic insect feeding monitors. Entomol Soc Am, Lanham, MD

McLean DL, Kinsey MG (1964) A technique for electronically recording aphid feeding and salivation. Nature 202:1358–1359

Tjallingii WF (1978) Electrical recording of aphid penetration. Entomol Exp Appl 24:521–530

Tjallingii WF (1985) Electrical nature of recorded signals during stylet penetration by aphids. Entomol Exp Appl 38:177–186

Walker GP, Backus EA (eds) (2000) Principles and applications of electronic monitoring and other techniques in the study of homepteran feeding behavior. Entomol Soc Am, Lanham, MD

Preface

The year was 2011. The event called "Hemipteran-Plant Interactions Symposium" was being held from July 11 to 14 at the University of São Paulo, Piracicaba Campus, which hosts the agronomy school "*Escola Superior de Agricultura Luiz de Queiroz*" (ESALQ), in São Paulo State, Brazil. Several scientists from around the world working with the electrical penetration graph or electropenetrography (EPG) were scheduled to speak, addressing their results on the use of EPG to monitor the feeding behavior of piercing-sucking insects. The talks on the order Hemiptera, though, were mostly on insects in the suborders Sternorrhyncha (scales, aphids, and white flies) and Auchenorrhyncha (leafhoppers, treehoppers, and planthoppers), with which the great majority of EPG work had been carried out. The only exception was the session organized by Paula Levin Mitchell (PLM), Winthrop University, USA, primarily on true bugs (suborder Heteroptera) and covering pioneering work on coreids, pentatomids, and mirids using EPG, which, at that time, was in its initial steps.

Antônio Ricardo Panizzi (ARP) was in the audience and approached PLM to learn more about the use of EPG in true bugs. One year later, in 2012, he imported his first EPG equipment (Giga-8) manufactured by Freddy Tjallingii from EPG Systems, Wageningen, the Netherlands. PLM visited his laboratory at Embrapa in Passo Fundo, RS, Brazil, and helped to install the EPG to start research work on true bugs. In 2013, ARP went to Fresno, California, to attend the International Workshop on Electrical Penetration Graph organized by Elaine A. Backus (EAB) and get trained on EPG; the heteropteran portion of that workshop was taught by PLM.

In this same year of 2013, a young entomologist named Tiago Lucini (TL) joined the lab to conduct his doctoral research, and from the very beginning, he got involved with EPG. Because of his great interest in EPG, arrangements were made to send him to California, to train on EPG work under the supervision of EAB at the USDA in Parlier, where he spent 6 months. After his return, a large chunk of data was generated on pentatomids, which culminated in the first published paper on stink bugs in 2016, highlighting waveforms and how to overcome the challenge of wiring pentatomids for EPG studies. Several other published papers followed with different species of stink bugs feeding on various crop plants (maize, soybean, and spring cereals).

In the meantime, new EPG equipment of the most recent generation, two four-channel AC-DC monitors, were acquired from EPG Technologies, Inc., Gainesville, FL, USA. As studies at the Embrapa lab continued, we (ARP and TL) came up with the idea to organize a workshop on EPG on pentatomids. This was finally conducted in 2019 – "First Workshop of EPG Studies on Pentatomids," at the Embrapa lab in Passo Fundo, which gathered participants from several countries. As a result of that, a technical bulletin was published (in Portuguese), and a national Brazilian network was established for EPG studies on true bugs. This aimed to have all interested researchers coordinated, to avoid research duplication, and to have a forum for general discussions on EPG issues on true bugs.

We (ARP and TL) then thought that it was time to gather all the scattered information on EPG studies on true bugs to be published in a book. We contacted PLM and invited her to join in this endeavor, which she promptly accepted. We approached Springer and negotiated how we could have our intended idea of a book on such a matter published. As the contract was signed, we started to work on this project.

The book contains eight chapters. In Chap. 1 a general overview is provided on phytophagous true bugs and electronic monitoring of their feeding behavior. Chapters 2 and 3 are devoted to heteropteran basic biology; in Chap. 2, mouthparts of true bugs are described and their feeding strategies are discussed, whereas Chap. 3 is dedicated to presenting and discussing the different feeding sites and resulting damage to plants. Chapters 4 and 5, respectively, provide detailed reviews of electronic monitoring of the feeding behavior of different species of stink bugs (Pentatomidae) and species of true bugs in all other families (Blissidae, Coreidae, Miridae, and Plataspidae) studied thus far. In Chap. 6, the general EPG procedures for true bugs (Heteroptera) are described; in Chap. 7, the role of EPG in developing and evaluating control methods for heteropteran crop pests is covered; and finally, in Chap. 8, our perspectives on the use of EPG in electronic monitoring of true bug activities – not only feeding, but all that can be investigated in the bug/plant/EPG interface – are presented and discussed.

One point of clarification: Often in the EPG literature, the term probing or probe is used to denote all the feeding activities of piercing-sucking insects from stylet insertion to stylet removal. The pros and cons of using this wording compared to the word feeding have been discussed in the literature (Tjallingii WF [1995] "Regulation of Phloem Sap Feeding by Aphids." In: Chapman RF, de Boer G (eds) *Regulatory Mechanisms in Insect Feeding*. Chapman & Hall, London, pp 190–209; Backus EA [2000] "Our Own Jabberwocky: Clarifying the Terminology of Certain Piercing-Sucking Behaviors of Homopterans." In: Walker GP, Backus EA (eds) *Principles and Applications of Electronic Monitoring and Other Techniques in the Study of Homopteran Feeding Behavior*. Entomological Society of America, Lanham, pp 1–13). Because to probe comes from the Latin word *probare*, which means test, inspect, show, demonstrate, we consider it less holistic and, therefore, less appropriate to denote activities related to the food intake process. Therefore, in this book, we decided to use the more general word feeding to denote stylet penetration and ingestion in a broad sense and probing specifically to denote food contact and brief penetration with the stylets to recognize food suitability.

We hope that this contribution on synthesizing EPG studies on true bugs will be of interest to all entomologists involved with electronic monitoring studies, and, in particular, to those studying the fascinating world of phytophagous true bugs.

Antônio Ricardo Panizzi Passo Fundo, RS, Brazil
Tiago Lucini
Paula Levin Mitchell Rock Hill, SC, USA

Acknowledgments

The publication of this book was only possible with the help of several persons to whom we want to express our sincerest gratitude. We thank João Pildervasser, Editor of Life Sciences and Neuroscience at Springer Nature (São Paulo, Brazil), who guided us during the publication process through the many steps taken toward the achievement of our goal. We are grateful to Sanjana Meenakshi Sundaram, Books Project Coordinator, and Vishnu Prakash of Springer Nature, SPi Global, Manapakkam, Chennai, India, for support during the handling of the book cover, proofs, indexes, and other details of the book production process. AR Panizzi thanks Embrapa Trigo (CNPT, Passo Fundo, RS, Brazil) and Conselho Nacional de Desenvolvimento Científico e Tecnológico (CNPq, Brasília, Brazil) for providing facilities and financial support when research with true bugs was conducted using the Electrical Penetration Graph (EPG), and when this book was under development (grant numbers 471517/2012-7; 301604/2013-4; 400551/2016-0; 302293/2017-5). He also thanks, the following postdocs, graduate students, undergraduate trainees, and lab assistants at the Universidade Federal do Paraná, Universidade de Passo Fundo, and Embrapa Trigo for supporting studies with the EPG: Tiago Lucini, Lisonéia F. Smaniotto, Alice Agostineto, Taynara Possebom, Vânia Bianchin, Maria Elaine Moreira Solagna, Elias do Amarante, and Odirlei Dalla Costa. T Lucini thanks CAPES (Coordenação de Aperfeiçoamento de Pessoal de Nível Superior, Brasília, Brazil) (grants 1561050/2013-0; 88887.371769/2019-0) and CNPq (grant 167371/2017-7) for financial support during his doctorate and postdoctorate studies with EPG on true bugs. PL Mitchell thanks Winthrop University for financial support during 20+ years of EPG work and all her former undergraduate and graduate research students at Winthrop, Clemson University, and the University of Delhi who have expanded our understanding of true bug feeding behavior, with special thanks to the Mitchell Lab EPG researchers: Amanda Leigh Dunlap, S. Bernell Cooke, Sarah Johnson-Lucas, Lisonéia F. Smaniotto, Francesca L. Stubbins, and Kelly LaRose Rivera. We also want to thank Elaine A. Backus (USDA/ARS, Parlier, CA, USA) for suggestions that much improved our chapters, and Simon L. Elliot, editor of Entomology in Focus, for the opportunity to have this book included in such a prestigious series.

Acknowledgements

Contents

About the Authors

Antônio Ricardo Panizzi received his Ph.D. from the University of Florida, USA, and works as a research entomologist at Embrapa in Brazil. He is an invited professor at the Federal University of Paraná, Brazil, and at the Universidad de la Republica in Uruguay. He received several awards, including *ESA International Branch Distinguished Scientist Award*, in Denver, CO, USA, 2017. He has served as editor of the *Annals of the Entomological Society of Brazil* presently Neotropical Entomology published by Springer. He is editor or co-editor of six books. He acted as head of the Entomology Team at Embrapa Soybean Research Center. Dr. Panizzi acted as consultant for soybean entomology programs in South America and for FAO in Turkey. He served as member of the Agronomy Consult Team for the National Science foundation of Brazil. Dr. Panizzi was invited scientist at the National Institute of Agro-Environmental Sciences, Tsukuba, Japan. He is honorary member and president of the Entomological Society of Brazil, and presently International Delegate of the Society. He has published 162 research articles and 49 book chapters.

Tiago Lucini received his doctoral degree from the Federal University of Paraná, Paraná, Brazil, in 2016. He has experience on the use of EPG (Electrical Penetration Graph) technique at the USDA in Parlier, California. He was an invited speaker at the XXV International Congress of Entomology at Orlando, FL, USA, and at the XXVII Brazilian Congress of Entomology at Gramado, RS, Brazil. He has developed postdoctorate research work on EPG at the National Wheat Research Center of Embrapa, in Passo Fundo, RS, Brazil, from 2016 to 2019. His current research interests focus on the bioecology and interactions of heteropterans with their wild and cultivated host plants, and the electronic monitoring of the feeding behavior of stink bugs. He has participated in pioneering publications on the use of EPG on phytophagous pentatomids, including abstracts in entomology congresses. Presently, he is research assistant at the Embrapa National Wheat Research Center.

Paula Levin Mitchell received her Ph.D. in zoology from The University of Texas at Austin (USA) in 1980 and subsequently held a 4-year postdoctoral appointment in entomology at Louisiana State University. After teaching for 25 years in the Biology Department at Winthrop University in South Carolina (USA), she retired as professor emerita. She continues to curate the insect collection at Winthrop University and is active in entomological educational outreach. Her specialty is coreid and pentatomid biology, with research interests encompassing feeding behavior, egg parasitoids, and the use of neem extracts to control heteropteran crop pests. She was a Fulbright senior research scholar in India and a visiting fellow at the University of Delhi and has served as co-editor of the *Journal of Economic Entomology*. Her publications include 29 research articles and 3 book chapters.

Chapter 1
Introduction to Electronic Monitoring of the Feeding Behavior of Phytophagous True Bugs

Contents

Abstract Electrical penetration graph technique, or electropenetrography (EPG), is a method of indirectly visualizing the feeding of piercing-sucking insects within plant tissue, permitting quantification of feeding behaviors such as stylet penetration, salivation, and ingestion from specific plant tissues. In this chapter, the basic principles of EPG are explained, and a brief history of development of the technique is presented, from early alternating current (AC) monitors, to direct current (DC) monitors, and finally to a combined AC-DC monitor that offers a choice of applied current and amplifier sensitivities. The signals, called waveforms, generated by EPG recordings are explained and related to the internal ingestion sites (e.g., xylem vessels, stem parenchyma). The two components of the waveforms, resistance to electron flow (R) and biological voltages (emf, or electromotive force), are distinguished. Biological information is reviewed for the five families of phytophagous Heteroptera that have been studied to date using EPG: Miridae (Cimicomorpha) and

A. R. Panizzi et al., *Electronic Monitoring of Feeding Behavior of Phytophagous True Bugs (Heteroptera)*, Entomology in Focus 6,
https://doi.org/10.1007/978-3-030-64674-5_1

Blissidae, Coreidae, Pentatomidae, and Plataspidae (Pentatomomorpha). Examples are provided of the role of EPG in characterizing damage from feeding by species of Pentatomidae, the most extensively studied family.

Keywords Electrical penetration graph · Electropenetrography · Waveforms · Feeding damage · Piercing-sucking insects

1.1 Introduction

Numerous species of phytophagous piercing-sucking insects from the suborder Heteroptera (Hemiptera) are economic pests with the potential to cause significant damage on various agricultural crops (Schaefer and Panizzi 2000; McPherson and McPherson 2000). Both vegetative and reproductive tissues are targeted by true bugs; the type and extent of damage vary with the feeding site and with the feeding strategy and tactics employed to extract nutrients (Hori 2000; Lucini and Panizzi 2018a).

Understanding the feeding behavior of heteropterans involves determining where the insect inserts its mouthparts (interlocked mandibles and maxillae, or stylets) and how long the stylets remain in each plant tissue, as well as feeding activities performed, such as initial and deep stylet penetration, salivation, and ingestion. With this information, it is possible to predict the extent of plant tissue damage and, most importantly, define appropriate control strategies for a specific arthropod pest.

The major obstacle to studying heteropteran feeding is the fact that their complex and highly sophisticated behaviors occur internally, within the plant tissue, which makes direct observation difficult (Walker 2000). However, in the 1960s a valuable technological advance was made with the introduction of an electronic system to monitor the interactions between piercing-sucking insects and their host plants (McLean and Kinsey 1964, 1965, 1967). In this technique, the sucking insect and its host plant compose part of a circuit in which an electrical current is applied to the insect-plant system, producing waveforms that represent each feeding activity (Tjallingii 1978; Walker 2000). This early Electronic Measurement System (EMS), later renamed the Electronic Monitoring System, applied only alternating current (AC). The technique was subsequently improved with the introduction of direct current (DC) (Tjallingii 1978), and eventually the option of either AC or DC (Backus and Bennett 2009), and was renamed EPG as an abbreviation for electrical penetration graph (Tjallingii 1985a) or electropenetrography (Backus et al. 2016).

Over the years, most electronic systems for monitoring feeding (including EPG) were designed primarily to study phytophagous sucking insects; however, some early studies using a different form of electronic monitoring were done in the 1970s with chewing insects (e.g., Kogan 1972, 1973). Nowadays, EPG with its advances and improvements, although still focused on herbivorous hemipterans, is also being used for other insects and arthropods, such as mites (Guo and Zhao 2000), herbivorous dipterans (Guedes et al. 2019), and bloodsuckers, such as mosquitoes (Wayadande et al. 2020) and ticks (Andrew Y. Li, unpublished data). Recently, the principles and applications of EPG were reviewed by Backus et al. (2019, 2021).

In the past 50+ years, the EPG technique has been widely applied to evaluate the behavioral activities of many species of piercing-sucking insects. Due to the worldwide spread of this technique plus its usefulness in different types of research, the number of published articles increased significantly in the last decade (Fig. 1.1a);

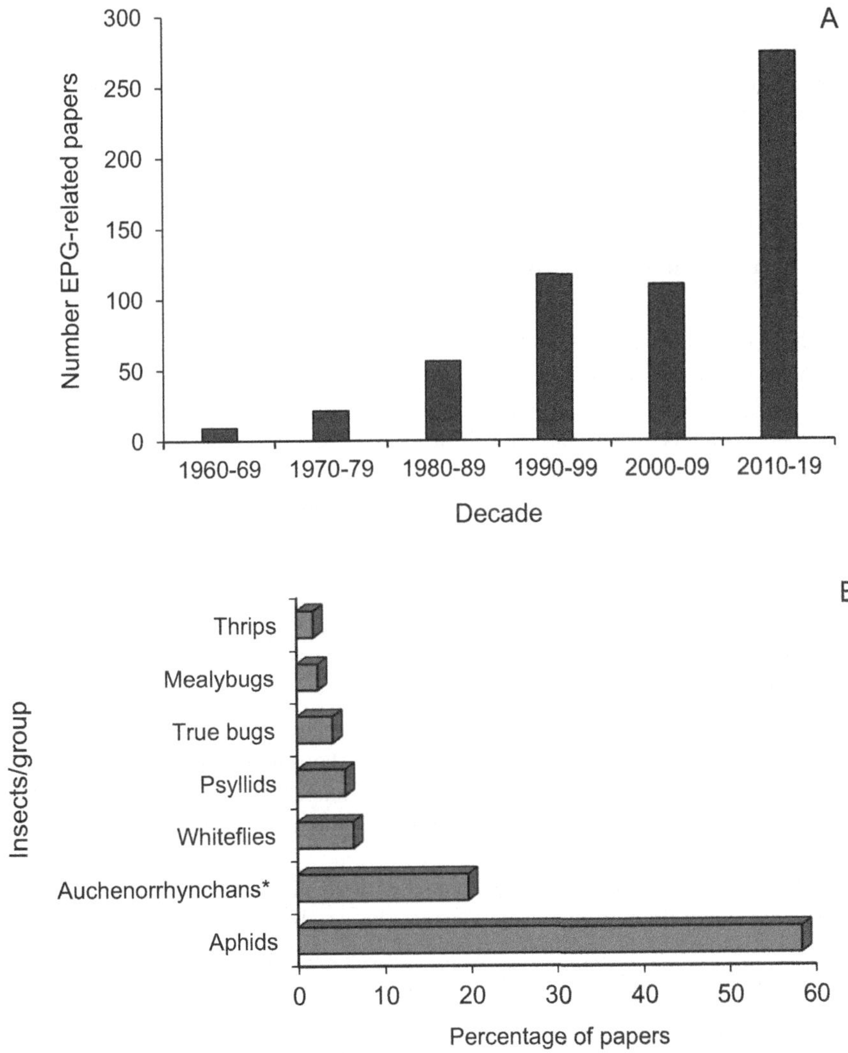

Fig. 1.1 Number of scientific articles published using the electrical penetration graph (EPG) technique to study the feeding behavior of phytophagous piercing-sucking insects over the decades since its invention (**a**); percentage of the total EPG articles published on different insect groups (**b**). *Includes leafhoppers and planthoppers. (Source: based on Scopus (2020); Web of Science (2020))

aphids and auchenorrhynchans are the most evaluated, constituting ca. 80% of the studies (Fig. 1.1b). Although this technology has been known for a long time and many species of heteropterans are recognized as crop pests worldwide, surprisingly little attention has been given to the use of EPG to study heteropteran feeding activities.

A literature review showed that 18 heteropteran species, from 5 different families, have been recorded using EPG to evaluate their feeding behavior, representing ca. 4% of the total scientific EPG articles published (Table 1.1; Fig. 1.1b). The first publication of EPG analysis of a true bug species appeared in 1990s with a coreid (Bonjour et al. 1991). Only recently the feeding behavior of a stink bug (Pentatomidae) was evaluated (Lucini and Panizzi 2016); eight pentatomid species have now been studied using EPG technology (see Sect. 1.4.4).

1.2 Principles of Electrical Penetration Graph Technique or Electropenetrography (EPG)

In this technique, the insect and its host plant make up part of a simple electrical circuit in which a low electrical current is applied and passes through the insect-plant interface. The electrical circuit consists of an output wire linked to a copper electrode, which is placed into the soil of a potted plant (or directly into a detached fruit or attached to a cutting in water) and an input wire, called the insect electrode. The insect electrode is composed of a brass nail with a copper electrode (± 3 cm long) soldered to it and a thin gold wire (~10–60 μm in diameter) glued to the copper electrode tip. The gold wire is then attached to the insect body (dorsal surface) using a conductive material (silver paint or silver glue). After that, the wiring stub with the bug attached is connected to the head stage amplifier in an AC-DC system (or to the EPG probe in a DC system) linked to the EPG monitor (Fig. 1.2) (Backus et al. 2019).

As the system is sensitive to electrical fluctuation, the presence of noise (external electrical signals captured from other places) in the output signal is common. To prevent such noise, the insects, plants, and head stage amplifiers are kept inside a Faraday cage in modern versions of this technology (Tjallingii 1978; Backus et al. 2016).

When the wired insect makes any contact with the plant, the circuit is completed; however, contact through tarsal movements (i.e., walking behavior) is limited. In contrast, when the insect inserts its stylets into the electrified plant tissues and ionized fluids move through the stylet food and salivary canals, the pathway becomes highly conductive. The insect acts as a variable resistor in the circuit, causing voltage in the system to fluctuate in recognizable patterns according to the activity performed (walking, stylet penetration, salivation, ingestion). These output signals, generated by the insect-plant interface, are captured and registered instantaneously as a waveform (Fig. 1.2) (Walker 2000; Backus et al. 2016). The changes in the

Table 1.1 Heteropteran species evaluated with EPG technology to monitor feeding behavior on their respective hosts

Species	Family	Stage used in EPG studies (instar/adult)	Host	References
Apolygus lucorum	Miridae	3rd, 4th	Green bean/ cotton/ *Helicoverpa armigera* eggs	Zhao et al. (2011), Li et al. (2016), and Lu et al. (2020)
Lygus hesperus		3rd, adults	Cotton/ other plants	Cline and Backus (2002), Backus et al. (2007), and Cervantes et al. (2016)
Lygus lineolaris		Adults	Cotton	Backus et al. (2018), Cervantes et al. (2016, 2017, 2019), and Tuelher et al. (2020)
Nesidiocoris tenuis		5th, adults	Tomato	Chinchilla-Ramírez et al. (2021)
Trigonotylus caelestialium		Adults	Wheat	Suzuki and Hori (2014) and Hori and Naito (2018)
Stenotus rubrovittatus		Adults	Wheat	Suzuki and Hori (2014) and Hori and Naito (2018)
Blissus insularis	Blissidae	Adults	St. Augustine grass	Backus et al. (2013) and Rangasamy et al. (2015)
Blissus occiduus		Adults	Buffalograss	Backus et al. (2013)
Anasa tristis	Coreidae	1st, 2nd, 4th and 5th	Cucurbit plants	Bonjour et al. (1991), Cook and Neal (1999), and Maskey (2010)[a]
Diceraeus (Dichelops) furcatus	Pentatomidae	Adults	Wheat	Lucini and Panizzi (2017b), and Lucini and Panizzi (2020)
Diceraeus (Dichelops) melacanthus		Adults	Maize	Lucini and Panizzi (2017a)
Edessa meditabunda		Adults	Soybean	Lucini and Panizzi (2016)
Euschistus heros		Adults	Soybean	Lucini and Panizzi (2018b)
Halyomorpha halys		5th	Broad bean	Serteyn et al. (2020a, b)
Nezara viridula		1st, 5th, adults	Soybean/ green bean	Mitchell et al. (2018) and Rivera and Mitchell (2020)
Piezodorus guildinii		Adults	Soybean	Lucini et al. (2016)

(continued)

Table 1.1 (continued)

Species	Family	Stage used in EPG studies (instar/adult)	Host	References
Tibraca limbativentris		Adults	Rice	Almeida et al. (2020)
Megacopta cribraria	Plataspidae	Adults	Soybean	Stubbins et al. (2017)

Source: adapted from Lucini and Panizzi (2018a)
[a]Master's thesis. It was not included in the calculations

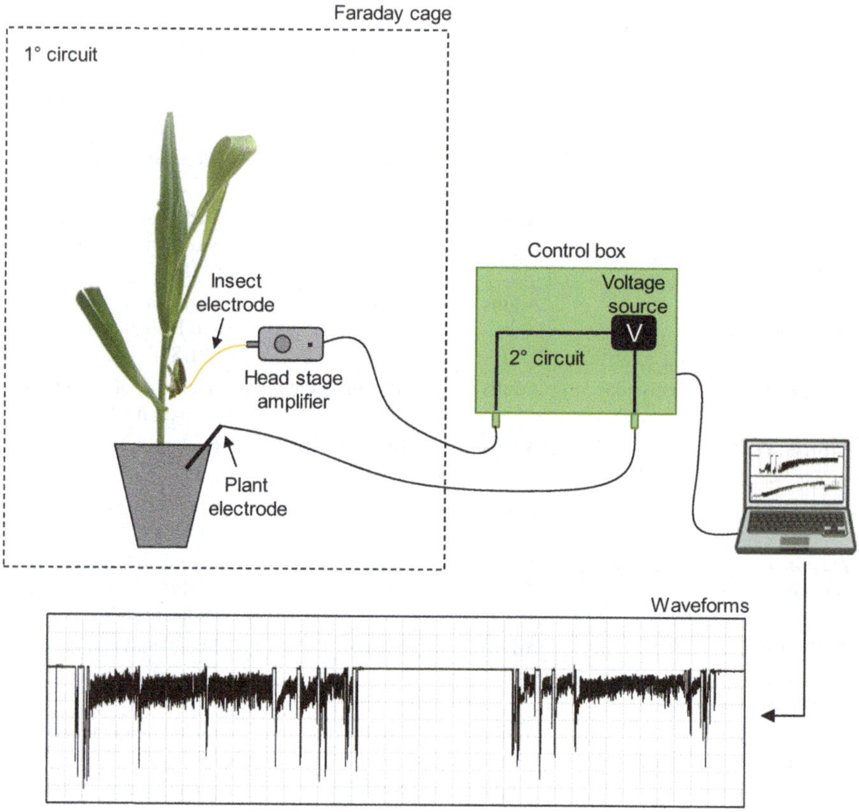

Fig. 1.2 Diagram of the electrical circuit of one electrical penetration graph channel showing its components (in an AC-DC monitor) and the waveforms obtained after recording. (Source: Lucini and Panizzi (2018a))

signal are directly related to the electrical origin or components of the waveforms: electrical resistances (R component) and biopotentials or biological voltages (emf component = electromotive force) (Tjallingii 1985a; Backus and Bennett 2009).

The R component is related to changes in electrical conductivity. This component may be affected by biological activities, such as the change in the ionic concentration of fluid in the salivary and food canals and the opening and closing of valves of the buccal cavity and the salivary pump (Tjallingii 1985a; Walker 2000). In the food canal, the ion (e.g., Na^+, K^+, Cl^-) concentration varies according to the type of fluid ingested; therefore, the ionic concentration directly affects the electrical resistance.

During the ingestion of fluids highly concentrated in ions, such as contents of parenchyma cells, the electrical charge flows without major restrictions, i.e., with low resistance (conductivity is high); thus, the R component is more emphasized. The opposite occurs in a fluid that is less concentrated, (e.g., xylem) which presents a high resistance (low conductivity). In the salivary canal, the concentration of ions varies according to the type and condition of saliva (gelling or watery). Gelling saliva, secreted to form a salivary sheath that lines the path of the stylets, is presumably more concentrated, so it has high conductivity in liquid form, but resistance increases as it solidifies. The more dilute watery saliva, rich in digestive enzymes, is presumable less conductive, but mixing of gelling and watery saliva during sheath formation enhances conductivity (Miles 1968, 1972; Walker 2000).

The emf component is related to the presence of biopotentials in the insect-plant interface, i.e., biological processes in the plant, insect, or both that generate small voltages that are added to the system. There are two major biological processes responsible for creating the emf component: (1) membrane potentials and (2) streaming potentials (potential = voltage). The membrane potential is triggered by the difference in electrical charges present on each side of the plant cell membrane. This results mainly from the difference in the concentration and type of ions (Na^+, Ca^{2+}, Cl^-, NO_3^-) on the inside and outside of the membrane. In general, on the inside portion of the membrane, the charges are negative, while on the outside they are positively charged (Fig. 1.3a), which generates a potential (voltage) ranging from ca. 100 to 180 mV (Walker 2000).

The existence of this potential is clear when recording the activities of small insects, such as aphids and whiteflies. During stylet penetration, these insects make occasional intracellular punctures (i.e., brief stylet penetration into a plant cell), while taking an intercellular path in the plant tissue. These punctures result in negative drops in voltage (Walker 2000), known as *potential drop* (pd-wave in DC recordings) (Fig. 1.3b) (Tjallingii 1985b). In the case of larger insects, such as stink bugs, there is no potential drop, because their stylets, much wider than those of aphids, take an intracellular path and simply break up the cells.

The streaming potential occurs when an ion-containing solution is forced to move between two ends of a capillary tube, such as the food canal (uptake) or salivary canal (secretion) of the maxillary stylets. During fluid movement in the tube, caused by hydrostatic pressure from the cibarial or salivary pumps, the positive or negative charges (ions) adhere to the capillary wall according to the physicochemical properties of the wall and the ions in the solution. In a narrow tube, the rate of flow of a solution is high in the center of the tube but decreases near the walls (Fig. 1.4). This causes a differential movement of ions and thereby creates a poten-

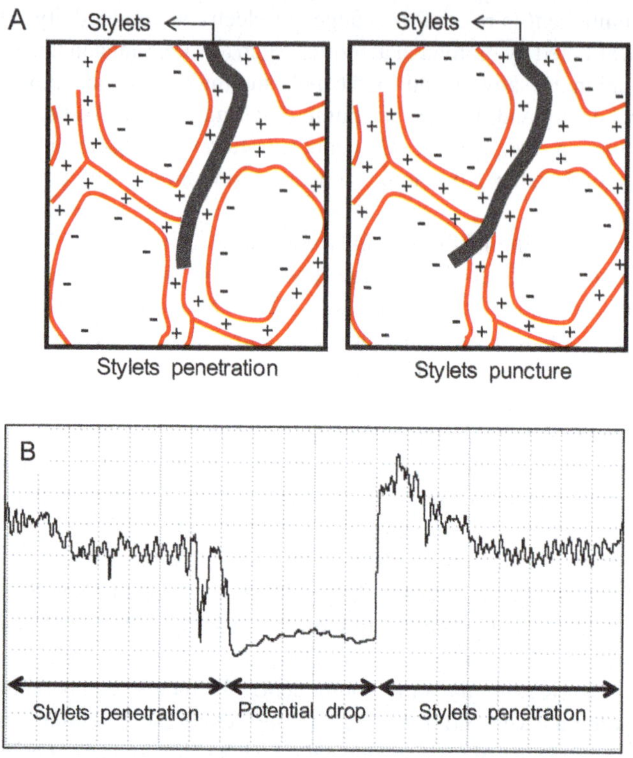

Fig. 1.3 Schematic representation of the inside and outside portions of the plant cell membrane, showing the arrangement of positive charges externally and negative charges internally; intercellular stylet pathway, left; stylets puncturing cell membrane, right (**a**). Aphid waveform showing a *potential drop* recorded at the moment that the stylets puncture the plant cell, causing a drop in voltage due to the presence of negative charges internally (**b**). (Sources: (**a**) adapted from Walker (2000); (**b**) T. Lucini (unpublished))

tial difference along the length of the tube; once fluid movement ceases, the streaming potential disappears (Walker 2000).

In EPG recordings, the equipment must be adjusted to capture both electrical components, R and emf, at a 50:50 ratio (or as close as possible), to obtain the maximum amount of information about the behavioral activities. Commonly, a given event and waveform may present both R and emf components simultaneously. Both components are additive; however, depending on the amplifier sensitivity of the primary circuit of the EPG system, one component is more emphasized than the other, and the relationship is reciprocal (Backus and Bennett 2009). Therefore, it is necessary to establish, via specific studies, the optimal level of amplifier sensitivity (also called input impedance or Ri level) to maximize both components.

The determination of R and emf components depends on two main factors: body size of the insect and input impedance applied. The relationship between the two components has been determined for aphids using empirical data (Tjallingii 1988).

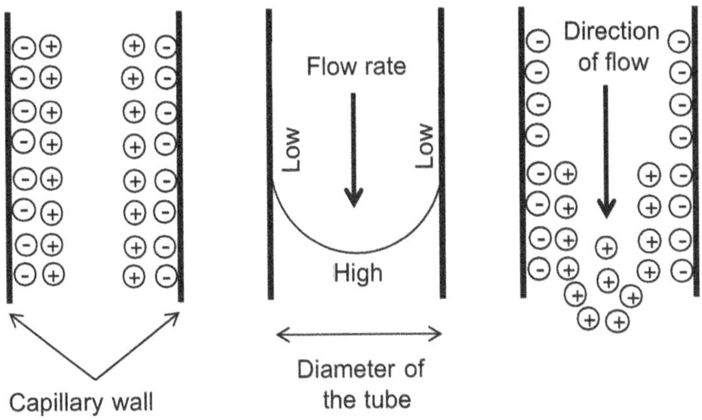

Fig. 1.4 Schematic representation of the occurrence of the streaming potential generated from the movement of fluids in a capillary tube, where the adhesion of charges (negative ions in this case) to the tube walls generates a charge separation and the uneven flow rate of fluid within the tube then creates the potential difference. (Source: Walker 2000)

According to Ohms' law, at low Ri levels the R component contributes more to the EPG signal, becoming only R at 10^6 Ohms for aphids. In contrast, the emf component is more predominant at higher Ri levels (> 10^9 Ohms), becoming exclusively emf at 10^{13} Ohms (Tjallingii 1988; Backus and Bennett 2009). For aphids, Tjallingii (1988) established that the best input impedance is 10^9 Ohms in the DC system, where both electrical components are captured in a similar proportion. In the case of heteropterans, preliminary studies have shown that, in general, a Ri from 10^7 Ohms to 10^8 Ohms achieves an optimal balance of both R and emf components of waveforms (Backus et al. 2013; Cervantes et al. 2016; Lucini and Panizzi 2018a). In general, data show that the bigger the insect (i.e., larger diameter of food and salivary canals, causing a greater movement of fluids, and, consequently, greater ionic conductivity), the lower the impedance level applied.

1.3 History of the Development of Electronic Monitoring and Comparison of Techniques

Over the last 50 years, three major generations of EPG devices have been designed: (1) the AC-EPG developed by McLean and Kinsey (1964), the DC-EPG developed by Tjallingii (1978), and the "universal" AC-DC EPG developed by Backus and Bennett (2009). Although each monitor employs different processing systems, the basic principle remains the same. A short history of each kind of monitor is presented and discussed below.

1.3.1 AC Monitor

The first generation of a monitor used to record the feeding behavior of a sucking insect [in this case the pea aphid – *Acyrthosiphon pisum* (Harris)] in real time was developed by McLean and Kinsey in the early 1960s, inspired by radio technology; the waveforms were recorded on a strip chart recorder (McLean and Kinsey 1964, 1965, 1967). This system was designed to use an AC applied signal and a fixed, low input impedance of 10^6 Ohms (McLean and Weigt 1968), and it was later named EMS (Electronic Measurement/ Monitoring System). In these monitors, the output signals captured and recorded by the equipment were mostly electrical resistance (R component) caused by the ionic charges of the fluids that moved through the stylets, such as saliva or cell contents from parenchyma or vascular tissues (Walker 2000).

Derived from the original device, other monitors were designed over the years with significant modifications (e.g., Brown and Holbrook 1976; Kinsey and McLean 1987; Backus and Bennett 1992), one of which is the device known as the "Missouri monitor," developed by Backus and Bennett (1992). Although those AC monitors presented significant differences among them, they are all commonly referred as AC systems, because they share a common principle: the applied voltage in the measuring circuit is based on amplitude modulation (AM), similar to radio circuits (Tjallingii 2000).

Using AC monitors, some crucial events performed during feeding were captured and recorded for the first time. For example, the initial and deep stylet penetration into the plant tissue, secretion of saliva, formation of a salivary sheath (pathway activities), as well as the termination of a feeding event (i.e., stylet withdrawal from plant tissue), and the ingestion of plant cell contents, such as from the vascular system, were all identified as waveforms (Backus et al. 2005a).

The early AC monitors measured fluctuations in the resistance component in the primary circuit, but were not sensitive enough to capture the emf components. This failure was assumed to be related to the low input impedance applied (10^6 Ohms), where the signal processing system substantially reduces/excludes the emf components from analysis (Backus et al. 2000; Tjallingii 2000). However, Backus and Bennett (2009) subsequently designed a more sophisticated AC monitor which was able to generate waveforms with detailed characteristics similar to those obtained using the DC system, including the emf component (see Sect. 1.3.3).

1.3.2 DC Monitor

Later in the 1970s, the second generation of EPG monitors was designed and developed using DC applied signal (commonly referred as the DC system). The first simple system to supply DC voltage was invented by Schaefers (1966). It was battery-operated, applied an input impedance of 10^6 Ohms, and was used until the 1980s. After that, Tjallingii (1978) (Wageningen University, The Netherlands) modified the system and published the first paper describing the improvements made to the system to record more detailed signals; the waveforms were recorded on a strip chart recorder (Tjallingii 1978).

A relevant modification was the introduction of a higher Ri level, where the system was changed to use an input impedance of either 10^9 or 10^{13} Ohms (Tjallingii 1978, 1985a, 1988). In addition, Tjallingii was responsible for introducing the concept of electrical origins of the waveforms (R and emf components), as previously discussed, which are essential to interpret the biological meanings of the waveforms (Walker 2000). With these upgrades, the optimized monitor recorded for the first time, beyond the R component, the intrinsic emf component present in the insect-plant interaction. The best-known DC monitor is the Giga model (Tjallingii 1978, 1988), which is still being manufactured. The term EPG was introduced for the technique at this time as an abbreviation for electrical penetration graph (Tjallingii 1985a) or, some years later, electropenetrography (Backus et al. 2016).

Originally, the DC system appeared to be more sensitive to external noise than the AC system, because the noise seemed to show a behavior similar to an emf component, which is captured by DC monitors but not by the early AC monitors (Walker 2000). However, the new AC-DC monitor (see Sect. 1.3.3) is also sensitive to the emf component; thus, the noise is also picked up during a recording using AC voltage on that monitor.

Improvement of the recording system, using computerized analog-to-digital (A/D) display to graph waveforms from both AC and DC systems, allowed the monitors to show waveforms in greater detail and clarity compared to previous strip chart recordings (Tjallingii 1995). Researchers commonly used AC-EPGs to record the feeding behavior of insects of medium-large body size (auchenorrhynchans, such as leafhoppers, and some heteropterans), whereas DC-EPGs were used for small insects (i.e., sternorrhynchans, such as aphids, whiteflies, and psyllids) (Backus 1994).

As AC monitors were no longer being manufactured, some researchers in the early years of this century used the DC monitor to study auchenorrhynchans (e.g., Lett et al. 2001; Stafford and Walker 2009; Miranda et al. 2009). However, recent studies have shown that the DC applied signal affects the feeding behavior of large auchenorrhynchans, such as the sharpshooter *Graphocephala atropunctata* (Signoret); the DC signal seems to reduce ingestion periods, as well as producing unusual waveforms more often compared to AC applied signal (Cervantes and Backus 2018). In addition, Backus et al. (2018) showed that the feeding behavior of the heteropteran *Lygus lineolaris* (Palisot de Beauvois) showed significant differences between AC and DC applied signals.

1.3.3 AC-DC Monitor

In order to develop a universal monitor that combined both AC and DC systems in the same equipment (third generation of EPGs), Tjallingii (2000) proposed the first AC-DC EPG system. This monitor was used to evaluate the feeding behavior of the thrips *Frankliniella occidentalis* Pergande (Kindt et al. 2006); however, the authors observed few changes in the information obtained and considered this monitor unsuccessful.

A second AC-DC monitor was proposed and developed by Backus and Bennett (2009) that combined features of the "Missouri" AC monitor and the DC monitor from Tjallingii, ensuring more flexibility in settings to monitor the feeding behavior of a variety of insects. Using the AC-DC monitor to record the feeding behavior of the pea aphid, *A. pisum*, the authors observed that the monitor graphed identical waveforms with high detail using either AC or DC applied signal; in addition, they found that using AC signal the monitor was able to register the emf component of the waveforms.

In addition to offering a choice of AC or DC signals, this new monitor was designed to use a variable and selectable range of input impedances (Ri levels) ranging from 10^6 to 10^{10} Ohms plus 10^{13} Ohms, which facilitates recording the feeding behavior of all types and sizes of insects. Further, the switchable input impedances make determination of the electrical components of waveforms (R and emf) faster and easier (Backus and Bennett 2009). The selectable Ri levels allow users to characterize and to create a waveform library, which is a set of waveforms produced by an insect recorded using the EPG at multiple input impedances (Backus et al. 2013; Pearson et al. 2014; Cervantes et al. 2016; Lucini et al. 2016; Lucini and Panizzi 2018b; Almeida et al. 2020).

The great flexibility of this new monitor is essential to users working on insects for which little or no previous information is available on feeding behavior; the optimal monitor adjustment to obtain the most information on the feeding process can be determined easily. For example, for stink bugs the best configuration of the monitor consists of applying a Ri of 10^7 Ohms and a low AC voltage, ca. 50 mV (Lucini and Panizzi 2018a).

1.3.4 EMIF Technique

The newest electronically monitored insect feeding (EMIF) technique (not to be confused with the AC monitor, which was sometimes called EMIF also) documents the feeding behavior of hemipterans but is less rigorous and precise than EPG. The system allows determination of the feeding frequency and duration of feeding activities over time, but unlike EPG, no waveforms are recorded. Shearer and Jones (1996) used an EMIF in environmental chambers to study the feeding behavior of *Nezara viridula* (L.).

Recently, Wiman et al. (2014) used a modified EMIF system from Shearer and Jones (1996) to monitor the feeding behavior of the brown marmorated stink bug *Halyomorpha halys* (Stål) on green bean pods. With this system design, the insect is not connected to a gold wire; instead, it is confined in a small box, called a "feeding chamber," composed of a conductive metal screen attached to a power source, to which is applied a DC current. The food to be tested, which is connected to the recording device with a wire, is positioned below the metal screen without touching it, but is close enough to allow insects to reach the food. The circuit is closed when the insect touches the food with its rostrum while standing on the screen, and then the current passes through the system until it returns to the recording instrument (schematic representation in Shearer and Jones 1996 and Wiman et al. 2014).

One improvement by Wiman and collaborators, compared with Shearer and Jones' design, is that they created "feeding tables," each one containing eight "feeding chambers," which allow the experimenter to record the feeding behavior of eight insects simultaneously in an environmental chamber (schematic representation and further details in Wiman et al. 2014). The primary advantage of this system compared with EPG is that large numbers of insects can be recorded simultaneously. However, the primary disadvantage is that there is no distinction between labial contacts with the surface of the food and actual stylet penetrations.

1.4 True Bug Species Studied Using EPG

1.4.1 Blissidae

The family Blissidae includes about 500 species in 50 genera with a worldwide distribution (Slater 1979). Before 1997, the group was considered a subfamily (Blissinae) of the family Lygaeidae; after that, it was raised to family status (Henry 1997). Blissids, known as chinch bugs, are phytophagous insects with a preference for monocotyledonous, mostly gramineous, plants (Poaceae) where they feed in the vascular system (Slater 1976, 1979). Bugs of the genus *Blissus* have been reported as pests for a long time on cultivated grasses in the United States (Reinert and Kerr 1973). Gramineous plants attacked by blissids include grain crops (e.g., maize, rice, wheat, and barley), grasses for livestock and dairy production, grasses used in athletic fields, golf courses, and others (references in Sweet 2000).

Adults of two congeneric species of Blissidae were recorded using EPG on different turf grasses, the southern chinch bug *Blissus insularis* Barber and the western chinch bug *Blissus occiduus* Barber (Backus et al. 2013).

1.4.2 Coreidae

The family Coreidae comprises 2,567 species in 436 genera (Schuh and Weirauch 2020) spread across all regions of the world, with greater abundance in the tropics (Schuh and Slater 1995). Coreids are commonly referred to as leaf-footed bugs, but they are also known as squash bugs and leatherbugs. In general, they present a medium to large body size (reaching 45 mm), representing some of the largest heteropterans; some species have a leaf-shaped expansion of the posterior tibia (Schuh and Slater 1995). They are plant feeders, and some species are considered pests on agricultural crops such as legumes, fruits, and garden vegetables (Mitchell 2000). Only nymphs of the species *Anasa tristis* (De Geer) have had their feeding behavior evaluated using EPG, on wild and cultivated cucurbit hosts (Bonjour et al. 1991; Cook and Neal 1999); the former was the first published EPG study for any heteropteran species.

1.4.3 Miridae

The family Miridae is the largest and most diverse family within Heteroptera comprising over 11,000 described species in more than 1300 genera; they are spread throughout all zoogeographic regions in the world, except Antarctica (Schuh and Slater 1995; Cassis and Schuh 2012; Schuh 2013). Mirids are, in general, small insects (1–15 mm), oligophagous, and commonly known as plant bugs (USA and Canada). Many species are economically important, attacking and damaging a variety of agricultural crops worldwide; however, a number of species are predatory and play an important role in biological control programs (Wheeler 2000a,b; Cassis and Schuh 2012).

Feeding behavior of nymphs and adults of six different species has been studied via EPG on different hosts, as follows: the green mirid bug *Apolygus lucorum* (Meyer-Dür), the tarnished plant bugs *Lygus hesperus* Knight, and *L. lineolaris* (Palisot de Beauvois), the tomato bug *Nesidiocoris tenuis* (Reuter) (Chinchilla-Ramírez et al. 2021), the sorghum plant bug *Stenotus rubrovittatus* (Matsumura), and the rice leaf bug *Trigonotylus caelestialium* (Kirkaldy) (all references in Table 1.1). All species were monitored exclusively on plants (including cotton, wheat, tomato, and others) except *A. lucorum*, which was recorded on green bean, cotton, and *Helicoverpa armigera* (Hübner) eggs. *A. lucorum* is known to act as an herbivore on several plants but also as a carnivore, preying on both insects and mites (Yuan et al. 2013). A number of mirid species are reported to be similarly zoophytophagous (Wheeler 2000b).

1.4.4 Pentatomidae

The family Pentatomidae is the fourth largest family within Heteroptera, comprising more than 4,700 species distributed in nearly 800 genera; they are found in all zoogeographic regions of the world (except Antarctica) with greater abundance in the tropical and subtropical regions (Grazia et al. 2015). Pentatomids range from 4 to 20 mm in body size (Schuh and Slater 1995) and are commonly called stink bugs. They are separated into nine subfamilies, of which seven are known to be phytophagous, feeding on cultivated and non-cultivated plants. The subfamily Asopinae comprises exclusively predacious insects; nothing is known of the habits of the subfamily Stirotarsinae (Schuh and Weirauch 2020).

Polyphagous plant feeding species in the subfamilies Edessinae and Pentatominae are considered key pests on many agricultural crops in the world, with a high potential to cause damage (Panizzi et al. 2000; Grazia et al. 2015). The genus *Edessa* is the most important in the subfamily Edessinae. The Pentatominae includes several genera with economic importance, such as *Aelia*, *Arvelius*, *Chinavia*, *Diceraeus* (*Dichelops*), *Euschistus*, *Mormidea*, *Nezara*, *Oebalus*, *Piezodorus*, *Plautia*, and *Tibraca* (Schuh and Slater 1995; Panizzi et al. 2000).

More pentatomids have been studied with EPG (eight species) than any other heteropteran family: the brown-winged stink bug, *Edessa meditabunda* (F.), which was the first EPG published for pentatomids (Lucini and Panizzi 2016); the red-banded stink bug, *Piezodorus guildinii* Westwood (Lucini et al. 2016); the so-called (in Brazil) green-belly stink bugs, *Diceraeus* (*Dichelops*) *melacanthus* (Dallas) (Lucini and Panizzi 2017a) and *Diceraeus* (*Dichelops*) *furcatus* (F.) (Lucini and Panizzi 2017b); the Southern green stink bug, or green vegetable bug, *N. viridula* (Mitchell et al. 2018; Rivera and Mitchell 2020); the Neotropical brown stink bug *Euschistus heros* (F.) (Lucini and Panizzi 2018b); the brown marmorated stink bug *H. halys* (Serteyn et al. (2020a,b), and the rice stalk stink bug *Tibraca limbativentris* Stål (Almeida et al. 2020).

1.4.5 Plataspidae

The family Plataspidae is represented by bugs that resemble beetles, with small to medium body size (2–20 mm); this family includes more than 500 species in about 60 genera (Henry 2009). Plataspids feed mainly on legumes (Schaefer 1988), and they are primarily restricted to the tropical regions of the Eastern Hemisphere (Schuh and Slater 1995). However, the species *Megacopta cribraria* (F.), known as the kudzu bug, has invaded the United States, where it became established as a pest of soybeans (Seiter et al. 2013). Until detection of *M. cribraria* in the USA, no other report of a plataspid bug had been made in the Western Hemisphere. The feeding behavior of adults of the invasive *M. cribraria* was recently characterized via EPG on soybean plants in the USA (Stubbins et al. 2017).

1.5 Feeding Sites and Relationship to Waveforms Generated Using EPG

Heteropterans exploit a wide variety of feeding sites on different plant structures during their feeding activities, including vegetative (stems and leaves), and reproductive structures (fruits and seeds). Each heteropteran group has a preference for one or more structures; for example, pentatomids have, in general, a strong preference for immature seeds (Schuh and Slater 1995), because the reserve tissue of the seeds (endosperm) contains essential nutrients, such as carbohydrates, lipids, and proteins, easily reached by their stylets (Slansky and Panizzi 1987).

During feeding, hemipterans perform distinct stylet activities in the tissue, of which three are highlighted here: (1) initial stylet insertion and deep penetration into plant tissue (named pathway activities); (2) salivation (including salivary sheath secretion, and pre-feeding activity); and (3) ingestion (fluid uptake) (Tjallingii 1978; Walker 2000). Each activity performed by the stylets inside the plant tissue is characterized by one or more distinct waveform types and/or subtypes. These are arranged for convenience into waveform families that reflect the underlying behavior (pathway, salivation, or ingestion) and are traditionally named using a combination of letters and numbers (e.g., I for ingestion, G1, G2 for pathway). However, pentatomid waveforms are named using a different convention established by Lucini and Panizzi (2016). For stylet activities within the plant tissue, the first letter of the genus and the species name is followed by a numeral indicating each waveform, e.g., Em1 for pathway waveforms of *E. meditabunda*. Naming of other pentatomid waveforms follows the traditional pattern established for most hemipterans, in which the letters Z or Np are used for non-penetrating activities. The waveform types/subtypes are described according to their shape and their electrical components: frequency (Hz), relative and absolute amplitude (%), voltage level (intra- or extracellular, most important for small insects) and electrical origin (R and emf components) (Tjallingii 1978, 1985b). The electrical origin is extremely important to determine the biological meaning of each waveform (Tjallingii 1978; Walker 2000).

The initial insertion of the stylets into the plant tissue is followed by deep stylet penetration to seek their ingestion sites. During this activity, the bugs may secrete gelling saliva to create a salivary sheath, which may be complete (i.e., reaching the target ingestion tissue) or incomplete, depending on the feeding strategy employed and the feeding site (see Chaps. 2 and 3; also Lucini and Panizzi 2018a and references therein). For true bugs studied thus far with the EPG technique, the feeding sites exploited include xylem vessels; phloem sieve elements; seed reserve tissue; leaf, stem, and fruit parenchyma; and staminate tissue of flower buds. Each feeding site is represented by characteristic waveform patterns with specific electrical components. In pentatomids, for example, the xylem and phloem waveforms are different from each other; whereas, the waveforms recorded from parenchyma tissue of vegetative structures and from seed reserve tissue are, in general, similar in appearance and electrical characteristics (Lucini and Panizzi 2018a).

EPG does not determine the position (ingestion tissue) of the stylets within plant tissue, it just records the pattern that represents that location and allows a biological meaning of the waveforms to be proposed. Therefore, additional studies are needed to correlate the waveforms registered with the position of the stylets in the plant tissues, i.e., to determine the location exploited by the bug during each waveform. The most frequently applied method is histological sectioning, by which the path taken by the stylets may be traced through plant tissue to determine the location where they terminate. This determination may be based on the terminal position of the salivary sheath (e.g., Mitchell et al. 2018) and/or based on the terminal position of the stylet tips, after being severed (e.g., Lucini and Panizzi 2016).

1.6 Damage to Plants Resulting from Feeding Activity

The damage caused by heteropterans is strongly correlated with the feeding strategies and tactics employed, which in turn are associated with differences in mouthpart morphology and salivary gland production (see Chap. 2 for more details). In general, visible and severe damage is observed when the bugs use the cell rupture feeding strategy to break up a pocket of plant cells mechanically (stylet action) and/or chemically (secretion of digestive enzymes) for subsequent ingestion of the cell contents (Backus et al. 2005b). The resulting damage includes several kinds of lesions, beginning with a local lesion at the stylet insertion point. This may lead to physiological disorders and occurrence of secondary symptoms from stylet insertion and secretion of saliva, such as tissue wilting and necrosis, deformation of leaves, fruits, and seeds, and abscission of reproductive structures (Hori 2000).

The sequence of damage caused by an adult of the pentatomid *D. melacanthus* after feeding for 8 h on a maize seedling is illustrated as an example (Fig. 1.5). After 24 h, small whitish regions (indicated by arrows) may be observed in the leaves where the bug fed, leading subsequently to more severe damage to the plant tissue.

Fig. 1.5 Progression of the damage (indicated by arrows) caused by *Diceraeus (Dichelops) melacanthus* adult on maize seedling (V2 stage) after feeding for 8 h under laboratory conditions. (Photos: T. Lucini)

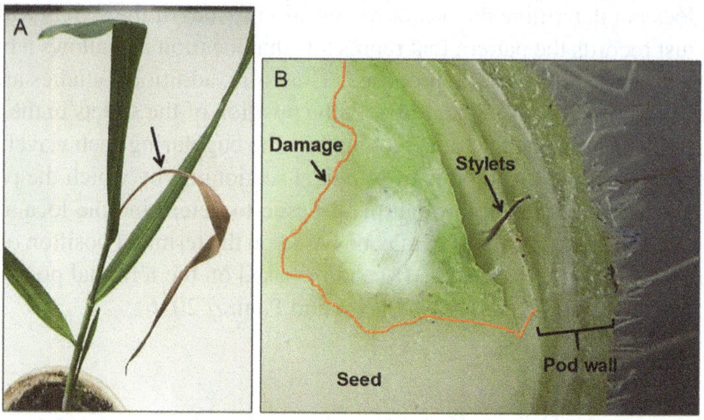

Fig. 1.6 Death of the central leaf of a wheat seedling (indicated by arrow), a symptom known as "dead heart," caused by the feeding activities of *Diceraeus* (*Dichelops*) *furcatus* (**a**); and damage caused by the stink bug *Piezodorus guildinii* on immature soybean seed (**b**). (Source: Fig. A = T. Lucini; B = Lucini et al. (2016))

Wilting and necrosis of the attacked regions are evident after 48 and 72 h. EPG monitoring revealed that the insect was using the cell rupture feeding strategy.

In more severe attacks, as observed in the field, extensive damage may occur in seedlings of crop plants such as wheat, including so-called dead heart (Fig. 1.6a; indicated by arrow). The central leaf of the seedling dies due to blockage at the attacked point, interrupting the flow of sap. On immature seeds, the primary feeding site for many species, stink bugs also use the cell rupture strategy to obtain nutrients from the reserve tissue, and, again, severe and visible damage is easily observed (Fig. 1.6b).

1.7 Concluding Remarks

For chewing insects, feeding behavior may be easily visualized and determined, whereas for piercing-sucking insects this is problematic, especially for the most relevant activities, which occur internally in host tissues. Although an insect with stylets inserted is generally thought to be feeding, many other activities might be performed inside the host and to discern them is always a challenge.

Previously, evaluation of the complex behavioral activities of these insects was based on visual observations and indirect quantification after feeding had ceased. For pentatomids, for example, assessment of feeding was primarily based on observations and counts of the external deposits of gelling saliva (often called flanges or stylet sheaths) left behind on the plant surface after feeding (Bowling 1979, 1980). However, direct feeding observation in real time and precise quantification could not be achieved until the development of a specialized technology, electronic monitoring of feeding behavior (later named EPG) of sucking insects. It was initially used in the 1960s with small sucking insects (primarily aphids); over time, EPG was used for other insect groups, and currently, it is the most sensitive and precise method available to monitor the feeding activities of Hemiptera.

From the basic characterization of the feeding behavior of sucking insects (i.e., determination of feeding sites and respective waveforms), the EPG technique may be applied in numerous other studies, such as feeding activities on resistant host plants (Rangasamy et al. 2015; Todd et al. 2016; Baldin et al. 2018); feeding activities on plants with induced defenses (Serteyn et al. 2020b); the effect of transgenic plants on feeding behavior (Yin et al. 2010; Cervantes et al. 2019); the process of acquisition and inoculation of phytopathogens (Sandanayaka et al. 2014; Maluta et al. 2019), and the effect of chemical compounds on feeding activities (Serikawa et al. 2012; Miranda et al. 2016; Lu et al. 2020). These are some possible EPG applications, and, as studies with EPG progress, much novel information will be generated, and more will be learned and added to what we presently know about the sophisticated feeding behavior of heteropterans.

References

Almeida ACS, Jesus FG, Barrigossi JAF (2020) Unveiling the feeding behavior of *Tibraca limbativentris* (Hemiptera: Pentatomidae) on rice using an electropenetrography waveform library. J Insect Sci 20:14;1–8

Backus EA (1994) History, development, and applications of the AC electronic monitoring system for insect feeding. In: Ellsbury MM, Backus EA, Ullman DL (eds) History, development, and application of AC electronic insect feeding monitors. Academic/Entomological Society of America, Lanham, pp 1–51

Backus EA, Bennett WH (1992) New AC electronic insect feeding monitor for fine-structure analysis of waveforms. Ann Entomol Soc Am 85:437–444

Backus EA, Bennett WH (2009) The AC-DC correlation monitor: new EPG design with flexible input resistors to detect both R and emf components for any piercing-sucking hemipteran. J Insect Physiol 55:869–884

Backus EA, Devaney MJ, Bennett WD (2000) Comparison of signal processing circuits among seven AC electronic monitoring systems for their effects on the emf and R components of aphids (Homoptera: Aphididae) waveforms. In: Walker GP, Backus EA (eds) Principles and applications of electronic monitoring and other techniques in the study of homopteran feeding behavior. Entomological Society of America, Lanham, pp 102–143

Backus EA, Habibi J, Yan F, Ellersieck M (2005a) Stylet penetration by adult *Homalodisca coagulata* on grape: electrical penetration graph waveform characterization, tissue correlation, and possible implications for transmission of *Xylella fastidiosa*. Ann Entomol Soc Am 98:787–813

Backus EA, Serrano MS, Ranger CM (2005b) Mechanisms of hopperburn: an overview of insect taxonomy, behavior, and physiology. Annu Rev Entomol 50:125–151

Backus EA, Cline AR, Ellerseick MR, Serrano MS (2007) *Lygus hesperus* (Hemiptera: Miridae) feeding on cotton: new methods and parameters for analysis of nonsequential electrical penetration graph data. Ann Entomol Soc Am 100:296–310

Backus EA, Rangasamy M, Stamm M, McAuslane HJ, Cherry R (2013) Waveform library for chinch bugs (Hemiptera: Heteroptera: Blissidae): characterization of electrical penetration graph waveforms at multiple input impedances. Ann Entomol Soc Am 106:524–539

Backus EA, Lin PA, Chang CJ, Shih HT (2016) Electropenetrography: a new diagnostic technology for study of feeding behavior of piercing-sucking insects. J Taiwan Agric Res 65:219–237

Backus EA, Cervantes FA, Godfrey L, Akbar W, Clark TL, Rojas MG (2018) Certain applied electrical signals during EPG cause negative effects on stylet probing behaviors by adult *Lygus lineolaris* (Hemiptera: Miridae). J Insect Physiol 105:64–75

Backus EA, Cervantes FA, Guedes RNC, Li AY, Wayadande AC (2019) AC–DC electropenetrography for in-depth studies of feeding and oviposition behaviors. Ann Entomol Soc Am 112:236–248

Backus EA, Guedes RNC, Reif KE (2021) AC-DC Electropenetrography: fundamentals, controversies, and perspectives for arthropod pest management. Pest Manage Sci 77:1132–1149

Baldin ELL, Stamm MD, Bentivenha JPF, Koch KG, Heng-Moss TM, Hunt TE (2018) Feeding behavior of *Aphis glycines* (Hemiptera: Aphididae) on soybeans exhibiting antibiosis, antixenosis, and tolerance resistance. Fla Entomol 101:223–228

Bonjour EL, Fargo WS, Webster JA, Richardson PE, Brusewitz GH (1991) Probing behavior comparisons of squash bugs (Heteroptera: Coreidae) on cucurbit hosts. Environ Entomol 20:143–149

Bowling CC (1979) The stylet sheath as an indicator of feeding activity of the rice stink bug. J Econ Entomol 72:259–260

Bowling CC (1980) The stylet sheath as an indicator of feeding activity by the southern green stink bug on soybeans. J Econ Entomol 73:1–3

Brown CM, Holbrook FR (1976) An improved electronics system for monitoring feeding of aphids. Am Potato J 53:457–462

Cassis G, Schuh RT (2012) Systematic, biodiversity, biogeography, and host associations of the Miridae (Insecta: Hemiptera: Heteroptera: Cimicomorpha). Annu Rev Entomol 57:377–404

Cervantes FA, Backus EA (2018) EPG waveform library for *Graphocephala atropunctata* (Hemiptera: Cicadellidae): effect of adhesive, input resistor, and voltage levels on waveform appearance and stylet probing behaviors. J Insect Physiol 109:21–40

Cervantes FA, Backus EA, Godfrey L, Akbar W, Clark TL (2016) Characterization of an EPG waveform library for adult *Lygus lineolaris* and *Lygus hesperus* (Hemiptera: Miridae) feeding on cotton squares. Ann Entomol Soc Am 109:684–697

Cervantes FA, Backus EA, Godfrey L, Wallis C, Akbar W, Clark TL, Rojas MG (2017) Correlation of electropenetrography waveforms from *Lygus lineolaris* (Hemiptera: Miridae) feeding on cotton squares with chemical evidence of inducible tannins. J Econ Entomol 110:2068–2075

Cervantes FA, Backus EA, Godfrey L, Rojas MG, Akbar W, Clark TL (2019) Quantitative differences in feeding behavior of *Lygus lineolaris* (Hemiptera: Miridae) on transgenic and nontransgenic cotton. J Econ Entomol 112:1920–1925

Chinchilla-Ramírez M, Garzo E, Fereres A, Gavara-Vidal J, ten Broeke CJM, van Loon JJA, Urbaneja A, Pérez-Hedo M (2021) Plant feeding by *Nesidiocoris tenuis*: quantifying its behavioral and mechanical components. Biol Control 152:104402;1–10

Cline AR, Backus EA (2002) Correlations among AC electronic monitoring waveforms, body postures, and stylet penetration behaviors of *Lygus hesperus* (Hemiptera: Miridae). Environ Entomol 31:538–549

Cook CA, Neal JJ (1999) Feeding behavior of larvae of *Anasa tristis* (Heteroptera: Coreidae) on pumpkin and cucumber. Environ Entomol 28:173–177

Grazia J, Panizzi AR, Greve C, Schwertner CF, Campos LA, Garbelotto TA, Fernandes JAM (2015) Stink bugs (Pentatomidae). In: Panizzi AR, Grazia J (eds) True bugs (Heteroptera) of the neotropics. Springer, Dordrecht, pp 681–756

Guedes RNC, Cervantes FA, Backus EA, Walse SS (2019) Substrate-mediated feeding and egg-laying by spotted wing drosophila: waveform recognition and quantification via electropenetrography. J Pest Sci 92:495–507

Guo F, Zhao Z (2000) Feeding behaviour of omethoate-resistant spider mites (Acari: Tetranychidae): a study using electrical penetration graphs. Syst Appl Acarol 5:3–7

Henry TJ (1997) Phylogenetic analysis of family groups within the infraorder Pentatomomorpha (Hemiptera: Heteroptera), with emphasis on the Lygaeoidea. Ann Entomol Soc Am 90:275–301

Henry TJ (2009) Biodiversity of the Heteroptera. In: Foottit RG, Adler PH (eds) Insect biodiversity: science and society. Wiley-Blackwell, Oxford, pp 223–263

Hori K (2000) Possible causes of disease symptoms resulting from the feeding of phytophagous Heteroptera. In: Schaefer CW, Panizzi AR (eds) Heteroptera of economic importance. CRC Press, Boca Raton, pp 11–36

Hori M, Naito S (2018) Feeding behaviors of rice ear bugs, *Trigonotylus caelestialium* and *Stenotus rubrovittatus* (Hemiptera: Miridae), in response to starch and its related substances. Appl Entomol Zool 53:143–150

Kindt F, Joostenb NN, Tjallingii WF (2006) Electrical penetration graphs of thrips revised: combining DC- and AC-EPG signals. J Insect Physiol 52:1–10

Kinsey RB, McLean DL (1987) Versatile electronic measurement system for studying probing and feeding behavior of piercing and sucking insects. Ann Entomol Soc Am 80:118–129

Kogan M (1972) Electronic recordings of feeding motions and rhythms of corn earworm (*Heliothis zea*) larvae. Isr J Entomol 7:49–58

Kogan M (1973) Automatic recordings of masticatory motions of leaf-chewing insects. Ann Entomol Soc Am 66:66–69

Lett JM, Granier M, Grondin M, Turpin P, Molinaro F, Chiroleu F, Peterschmitt M, Reynaud B (2001) Electrical penetration graphs from *Cicadulina mbila* on maize, the fine structure of its stylet pathways and consequences for virus transmission efficiency. Entomol Exp Appl 101:93–109

Li W, Wyckhuys KAG, Wu K (2016) Does feeding behavior of a zoophytophagous mirid differ between host plant and insect prey items? Arthropod Plant Interact 10:79–86

Lu Z, Dong S, Li C, Li L, Yu Y, Men X, Yin S (2020) Sublethal and transgenerational effects of dinotefuran on biological parameters and behavioural traits of the mirid bug *Apolygus lucorum*. Sci Rep 10:226;1–8

Lucini T, Panizzi AR (2016) Waveform characterization of the soybean stem feeder *Edessa meditabunda* (F.) (Hemiptera: Heteroptera: Pentatomidae): overcoming the challenge of wiring pentatomids for EPG. Entomol Exp Appl 158:118–132

Lucini T, Panizzi AR (2017a) Feeding behavior of the stink bug *Dichelops melacanthus* Dallas on corn seedlings: an EPG analysis at multiple input impedances and histology correlation. Ann Entomol Soc Am 110:160–171

Lucini T, Panizzi AR (2017b) Probing behavior of *Dichelops furcatus* (F.) (Heteroptera: Pentatomidae) on wheat plants characterized by electropenetrography (EPG) and histological studies. J Insect Sci 17:65;1–15

Lucini T, Panizzi AR (2018a) Electropenetrography (EPG): a breakthrough tool unveiling stink bug (Pentatomidae) feeding on plants. Neotrop Entomol 47:6–18

Lucini T, Panizzi AR (2018b) Electropenetrography monitoring of the neotropical brown-stink bug (Hemiptera: Pentatomidae) on soybean pods: an electrical penetration graph-histology analysis. J Insect Sci 18:5;1–14

Lucini T, Panizzi AR (2020) Electropenetrographic comparison of feeding behavior of *Dichelops furcatus* (Hemiptera: Heteroptera: Pentatomidae) on soybean and spring cereals. J Econ Entomol 113:1796–1803

Lucini T, Panizzi AR, Backus EA (2016) Characterization of an EPG waveform library for red-banded stink bug, *Piezodorus guildinii* (Hemiptera: Pentatomidae), on soybean plants. Ann Entomol Soc Am 109:198–210

Maluta N, Fereres A, Lopes JRS (2019) Plant-mediated indirect effects of two viruses with different transmission modes on *Bemisia tabaci* feeding behavior and fitness. J Pest Sci 92:405–416

Maskey K (2010) Comparison of electrical penetration graph waveforms of squash bug feeding on watermelon and its relatives. Master's thesis, Oklahoma State University, Stillwater, Oklahoma

McLean DL, Kinsey MG (1964) A technique for electronically recording aphid feeding and salivation. Nature 202:1358–1359

McLean DL, Kinsey MG (1965) Identification of electrically recorded curve patterns associated with aphid feeding and salivation. Nature 205:1130–1131

McLean DL, Kinsey MG (1967) Probing behavior of the pea aphid, *Acyrthosiphon pisum*. I. Definitive correlation of electronically recorded waveforms with aphid probing activities. Ann Entomol Soc Am 60:400–406

McLean DL, Weigt WA Jr (1968) An electronic measuring system to record aphid salivation and ingestion. Ann Entomol Soc Am 61:180–185

McPherson JE, McPherson RM (2000) Stink bugs of economic importance in America North of Mexico. CRC Press, Boca Raton

Miles PW (1968) Insect secretions in plants. Annu Rev Phytopathol 6:137–164

Miles PW (1972) The saliva of Hemiptera. Adv Insect Physiol 9:183–255

Miranda MP, Fereres A, Appezzato-Da-Gloria B, Lopes JRS (2009) Characterization of electrical penetration graphs of *Bucephalogonia xanthophis*, a vector of *Xylella fastidiosa* in citrus. Entomol Exp Appl 130:35–46

Miranda MP, Yamamoto PT, Garcia RB, Lopes JPA, Lopes JRS (2016) Thiamethoxam and imidacloprid drench applications on sweet orange nursery trees disrupt the feeding and settling behaviour of *Diaphorina citri* (Hemiptera: Liviidae). Pest Manag Sci 72:1785–1793

Mitchell PL (2000) Leaf-footed bugs (Coreidae). In: Schaefer CW, Panizzi AR (eds) Heteroptera of economic importance. CRC Press, Boca Raton, pp 337–403

Mitchell PL, Cooke SB, Smaniotto LF (2018) Probing behavior of *Nezara viridula* on soybean: characterization and comparison of electrical penetration graph (EPG) waveforms on vegetative and reproductive plant structures. J Agric Urban Entomol 34:19–43

Panizzi AR, McPherson JE, James DG, Javahery M, McPherson RM (2000) Stink bugs (Pentatomidae). In: Schaefer CW, Panizzi AR (eds) Heteroptera of economic importance. CRC Press, Boca Raton, pp 421–474

Pearson CC, Backus EA, Shugart HJ, Munyaneza JE (2014) Characterization and correlation of EPG waveforms of *Bactericera cockerelli* (Hemiptera: Triozidae): variability in waveform appearance in relation to applied signal. Ann Entomol Soc Am 107:650–666

Rangasamy M, Mcauslane HJ, Backus EA, Cherry RH (2015) Differential probing behavior of *Blissus insularis* (Hemiptera: Blissidae) on resistant and susceptible St. Augustinegrasses. J Econ Entomol 108:780–788

Reinert JA, Kerr SH (1973) Bionomics and control of lawn chinch bugs. Bull Entomol Soc Am 19:91–92

Rivera KL, Mitchell PL (2020) Feeding behavior of *Nezara viridula* first instars: EPG analysis and effect of food availability on subsequent development. J Agric Urban Entomol 36:47–63

Sandanayaka WRM, Moreno A, Tooman LK, Page-Weir NEM, Fereres A (2014) Stylet penetration activities linked to the acquisition and inoculation of *Candidatus liberibacter solanacearum* by its vector tomato potato psyllid. Entomol Exp Appl 151:170–181

Schaefer CW (1988) The food plants of some “primitive” Pentatomoidea (Hemiptera: Heteroptera). Phytophaga 2:19–45

Schaefer CW, Panizzi AR (2000) Heteroptera of economic importance. CRC Press, Boca Raton

Schaefers GA (1966) The use of direct current for electronically recording aphid feeding and salivation. Ann Entomol Soc Am 59:1022–1024

Schuh RT (2013) On-line systematic catalog of plant bugs (Insecta: Heteroptera: Miridae). Available in: http://research.amnh.org/pbi/catalog/ (2002–2013). Accessed 31 Oct 2019

Schuh RT, Slater JA (1995) True bugs of the world (Hemiptera: Heteroptera), classification and natural history. Cornell University Press, New York

Schuh RT, Weirauch C (2020) True bugs of the world (Hemiptera: Heteroptera): classification and natural history, 2nd edn. Siri Scientific Press, Manchester

Scopus. Available in: https://www.scopus.com. Accessed 16 Mar 2020

Seiter NJ, Greene JK, Reay-Jones FPF (2013) Reduction of soybean yield components by *Megacopta cribraria* (Hemiptera: Plataspidae). J Econ Entomol 106:1676–1683

Serikawa RH, Backus EA, Rogers ME (2012) Effects of soil-applied imidacloprid on Asian citrus psyllid (Hemiptera: Psyllidae) feeding behavior. J Econ Entomol 105:1492–1502

Serteyn L, Ponnet L, Backus EA, Francis F (2020a) Characterization of electropenetrography waveforms for the invasive heteropteran pest, *Halyomorpha halys*, on *Vicia faba* leaves. Arthropod Plant Interact 14:113–126

Serteyn L, Ponnet L, Saive M, Fauconnier ML, Francis F (2020b) Changes of feeding behavior and salivary proteome of brown marmorated stink bug when exposed to insect-induced plant defenses. Arthropod Plant Interact 14:101–112

Shearer PW, Jones VP (1996) Diel feeding pattern of adult female southern green stink bug (Hemiptera: Pentatomidae). Environ Entomol 25:599–602

Slansky F Jr, Panizzi AR (1987) Nutritional ecology of seed-sucking insects. In: Slansky F Jr, Rodriguez JG (eds) Nutritional ecology of insects, mites, spiders, and related invertebrates. Wiley, New York, pp 283–320

Slater JA (1976) Monocots and chinch bugs: a study of host plant relationships in the lygaeid subfamily Blissinae (Hemiptera: Lygaeidae). Biotropica 8:143–165

Slater JA (1979) The systematics, phylogeny, and zoogeography of the Blissinae of the world (Hemiptera, Lygaeidae). Bull Am Mus Nat Hist 165:1–180

Stafford CA, Walker GP (2009) Characterization and correlation of DC electrical penetration graph waveforms with feeding behavior of beet leafhopper, *Circulifer tenellus*. Entomol Exp Appl 130:113–129

Stubbins FL, Mitchell PL, Turnbull MW, Reay-Jones FPF, Greene JK (2017) Mouthpart morphology and feeding behavior of the invasive kudzu bug, *Megacopta cribraria* (Hemiptera: Plataspidae). Invert Biol 136:309–320

Suzuki Y, Hori M (2014) Diurnal locomotion and feeding activities of two rice-ear bugs, *Trigonotylus caelestialium* and *Stenotus rubrovittatus* (Hemiptera: Heteroptera: Miridae). Appl Entomol Zool 49:149–157

Sweet MH II (2000) Seed and chinch bugs (Lygaeoidea). In: Schaefer CW, Panizzi AR (eds) Heteroptera of economic importance. CRC Press, Boca Raton, pp 143–264

Tjallingii WF (1978) Electronic recording of penetration behaviour by aphids. Entomol Exp Appl 24:721–730

Tjallingii WF (1985a) Electrical nature of recorded signals during stylet penetration by aphids. Entomol Exp Appl 38:177–186

Tjallingii WF (1985b) Membrane potentials as an indication for plant cell penetration by aphid stylets. Entomol Exp Appl 38:187–193

Tjallingii WF (1988) Electrical recording of stylet penetration activities. In: Minks AK, Harrewijn P (eds) World crop pests: aphids, their biology, natural enemies and control. Elsevier, Amsterdam, pp 95–108

Tjallingii WF (1995) Regulation of phloem sap feeding by aphids. In: Chapman RF, de Boer G (eds) Regulatory mechanisms in insect feeding. Chapman & Hall, New York, pp 190–209

Tjallingii WF (2000) Comparison of AC and DC systems for electronic monitoring of stylet penetration activities by homopterans. In: Walker GP, Backus EA (eds) Principles and applications of electronic monitoring and other techniques in the study of homopteran feeding behavior. Entomological Society of America, Lanham, pp 41–69

Todd JC, Rouf Mian MA, Backus EA, Finer JJ, Redinbaugh MG (2016) Feeding behavior of soybean aphid (Hemiptera: Aphididae) biotype 2 on resistant and susceptible soybean. J Econ Entomol 109:426–433

Tuelher ES, Backus EA, Cervantes F, Oliveira EE (2020) Quantifying *Lygus lineolaris* stylet probing behavior and associated damage to cotton leaf terminals. J Pest Sci 93:663–677

Walker GP (2000) A beginner's guide to electronic monitoring of homopteran probing behavior. In: Walker GP, Backus EA (eds) Principles and applications of electronic monitoring and other techniques in the study of homopteran feeding behavior. Entomological Society of America, Lanham, pp 14–40

Wayadande AC, Backus EA, Noden BH, Ebert T (2020) Waveforms from stylet probing of the mosquito *Aedes aegypti* (Diptera: Culicidae) measured by AC–DC electropenetrography. J Med Entomol 57:353–368

Web of Science. Available in: https://www.webofknowledge.com. Accessed 16 Mar 2020

Wheeler AG (2000a) Plant bugs (Miridae) as plant pests. In: Schaefer CW, Panizzi AR (eds) Heteroptera of economic importance. CRC Press, Boca Raton, pp 37–83

Wheeler AG (2000b) Predacious plant bugs (Miridae). In: Schaefer CW, Panizzi AR (eds) Heteroptera of economic importance. CRC Press, Boca Raton, pp 657–693

Wiman NG, Walton VM, Shearer PW, Rondon SI (2014) Electronically monitored labial dabbing and stylet 'probing' behaviors of brown marmorated stink bug, *Halyomorpha halys*, in simulated environments. PloS One 9:e113514;1–24

Yin HD, Wang XY, Xue K, Huang CH, Wang RJ, Yan FM, Xu CR (2010) Impacts of transgenic Bt cotton on the stylet penetration behaviors of *Bemisia tabaci* biotype B: evidence from laboratory experiments. Insect Sci 17:344–352

Yuan W, Li WJ, Lu YH, Wu KM (2013) Combination of plant and insect eggs as food sources facilitates ovarian development in an omnivorous bug *Apolygus lucorum* (Hemiptera: Miridae). J Econ Entomol 106:1200–1208

Zhao QJ, Wu D, Lin FM, Li CY, Zhang YJ, Wu KM, Guo YY (2011) EPG analysis of *Apolygus lucorum* Meyer-Dür feeding behaviors on different cotton varieties (lines) and field verifications. Sci Agric Sin 44:2260–2268

Chapter 2
Mouthparts Description and Modes of Feeding of Phytophagous True Bugs

Contents

Abstract Heteropteran mouthparts consist of elongate mandibular and maxillary stylets encased at rest ventrally within the labial sheath. The greatly reduced labrum covers the stylets proximally as they emerge from the head. The two mandibular stylets surround the two maxillae, which are appressed. The food and salivary canals are formed by grooves and interlocking ridges in the maxillary stylets. Only the stylets enter the plant tissue; in phytophagous Heteroptera, all four stylets penetrate to an equal depth, with the maxillary stylets leading slightly. Details of the morphology and sensory capabilities of the mouthparts and foregut, the types of saliva, and the process of stylet penetration and ingestion are provided. The different modes of feeding of various heteropteran groups are explained, including salivary sheath, cell rupture, and osmotic pump, with a historical overview of the development of our current understanding of feeding strategies and tactics in true bugs.

Keywords Mouthparts · Stylet penetration · Feeding strategy · Salivary sheath · Cell rupture · Osmotic pump

A. R. Panizzi et al., *Electronic Monitoring of Feeding Behavior of Phytophagous True Bugs (Heteroptera)*, Entomology in Focus 6,
https://doi.org/10.1007/978-3-030-64674-5_2

2.1 Introduction

Mouthparts adapted for piercing and sucking allow precise placement of the stylets into specific internal plant cells or tissues. Lacking heavily sclerotized mandibles to chew leaves, stems, and roots, piercing-sucking insects rely on extraoral digestion by salivary enzymes or specialize on liquids flowing within the plant's vascular system. Within these constraints, various feeding modes have evolved in phytophagous Hemiptera to extract nutrients using the stylets. Of the three hemipteran suborders, Heteroptera present the greatest variety in feeding behavior, with large differences in stylet length, an array of potent salivary enzymes, and wide variation in preferred ingestion sites. In this chapter, we will provide detailed descriptions of heteropteran mouthparts and will track the development of our understanding of feeding modes in Heteroptera.

2.2 Mouthparts Description

Heteropteran mouthparts consist of the labrum, the labium, two mandibular stylets, and two interlocking maxillary stylets. Neither labial nor maxillary palps are present. The entire structure, excluding the labrum, is referred to as the rostrum and connects near the front of the head, unlike the attachment point in other Hemiptera. In most Heteroptera (and in all phytophagous groups), the labrum is reduced to a grooved flap or cone, articulating with the distal portion of the clypeus (i.e., the tylus, or anteclypeus) and covering the anterior portion of the stylet bundle while at rest (Awati 1914; Snodgrass 1935; J. F. Esquivel, cited in Mitchell et al. 2018a). The tube-like labium encases the stylets, when they are not in active use, within an open longitudinal groove located anteriorly (i.e., ventrally when in resting position against the abdomen). The stylets at rest do not extend beyond the tip of the labium. In all phytophagous species, the mandibular stylets surround the maxillary stylets, which interlock to form the separate salivary and food canals (Snodgrass 1935). Only the stylets penetrate the plant tissue, to deposit or inject saliva and extract plant nutrients.

2.2.1 The Labium

The heteropteran labium is divided into three or four segments. Articulation and flexibility of individual segments varies between and within infraorders, with an evolutionary progression toward decreased mobility of the apical segment (Cobben 1978, 1979). All phytophagous groups have a four-segmented labium (Schuh and Weirauch 2020) capable of elbow folding (Cobben 1978). The tip of the labium possesses both chemo- and mechanosensory sensilla (Backus 1988) which allow investigation of the plant before stylet insertion. Additional mechanoreceptors occur along the sides of the labium; these likely play a role in the movement of labial segments as the labium folds back during stylet penetration (see Sect. 2.2.5) (Backus 1988; Wang and Dai 2017). In tingids, pentatomomorphans, and most mirids, each labial segment has a pair of sutures running longitudinally that are lacking in other infraorders; this cuticular arrangement may permit compression of the labial segments as they form an elbow fold during feeding (Cobben 1978).

To some extent, length of the rostrum reflects overall body size; in phytophagous pentatomids with similar feeding habits, larger species had longer rostra (Depieri and Panizzi 2010a). However, in some fruit and seed-feeding species, labium length has evolved to match the depth of penetration necessary to reach developing or mature seeds within fruit. The Australian soapberry bug, *Leptocoris tagalicus* Hahn (Rhopalidae), evolved longer mouthparts to reach the seeds of the invasive balloon vine, *Cardiospermum grandiflorum* Swartz (Carroll et al. 2005). Similarly, in the southern United States, host races of a soapberry bug, *Jadera haematoloma* Herrich-Schäffer, have adapted to the fruits of different species of introduced sapindaceous hosts by altering rostral lengths (Carroll and Boyd 1992). These morphological changes, driven by directional selection, have taken place over just a few decades (Carroll et al. 2005).

In the coreid genus *Leptoglossus* (Guérin-Méneville), the ratio of adult rostral length to body length varies dramatically by species depending on host plant and diameter of the fruit, ranging from 0.70 in the magnolia bug, *L. fulvicornis* Westwood, to 0.39 in the mistletoe specialist, *L. brevirostris* Barber. In magnolia bug nymphs, the rostrum may be longer than the body (P. L. Mitchell, unpublished). A similarly elongated labium, especially in nymphs, is seen in *Hygia cliens* Dolling (Coreidae), which feeds on phloem deep within tree trunks (Maschwitz et al. 1987). Different mouthpart adaptations may be seen in other species whose food source is located at a considerable depth in the plant. Some plataspids, aradids, and cydnids have long stylets but a relatively short labium; the stylets at rest may be coiled within the head, in an internal pouch extending into the abdomen, or in a bulbous expansion of the second labial segment (Snodgrass 1935; Cobben 1978; Maschwitz et al. 1987).

The thickness and overall shape of the labium also varies. Predatory reduviids typically have a strongly curved, stout rostrum (Schuh and Weirauch 2020), whereas in plant feeding Pentatomomorpha, the rostrum is thin and straight. In the predatory Asopinae (Pentatomidae), the rostrum is markedly thickened, with the first labial segment particularly enlarged, compared with phytophagous pentatomids (Rider et al. 2018).

Sensory structures on the tip of the labium are similar among phytophagous bugs in the infraorders Cimicomorpha and Pentatomomorpha and include hairs (strictly mechanosensory) and pegs (chemosensory or mixed function) (Backus 1988). Labial dabbing, brief repeated contacts between the labial tip and the plant surface, may also be accompanied by exudation and re-intake of small quantities of watery saliva (Miles 1958, 1959). The saliva, having contacted the plant surface, is sucked back into the stylets and passes into the precibarium. Thus, labial dabbing permits gustatory exploration involving both the labial and precibarial sensilla prior to actual ingestion (Backus 1988). Recent detailed studies of the labial sensilla in *Erthesina fullo* (Thunberg) (Pentatomidae) revealed chemosensory pegs (sensilla basiconica) at the labial tip intermingled with a brush of long mechanosensory hairs (sensilla trichodea) (Wang and Dai 2020). The authors speculate that the hairs serve to smear watery saliva on the plant surface to facilitate tasting.

2.2.2 *The Stylet Bundle*

The stylet bundle is composed of two outer mandibular stylets and two interlocked maxillary stylets. The food canal and (when present) the salivary canal are both located between the conjoined maxillary stylets. The mandibular stylets surround and encase the maxillae and may be interlocked with the maxillae (Cobben 1978). In mirids, tingids, and all Pentatomomorpha, the entire bundle penetrates the plant tissue to the same depth, with the maxillary stylets leading slightly (Cobben 1978; Backus 1988). Coordinated movement of all four stylets within the host tissue, with the mandibles guiding the movement of the stylet bundle, is thought to be an evolutionarily advanced trait in Heteroptera and a prerequisite for phytophagy (Cobben 1979). In predatory infraorders such as Gerromorpha and Dipsocoromorpha, the mandibles serve to "harpoon" the prey cuticle, and the maxillary stylets move within the body of the prey, independent of the mandibles, to lacerate the tissues (Cobben 1978).

Stylets of Heteroptera possess only mechanosensory sensilla, used for proprioception. In Cimicomorpha, each of the four stylets is provided with a nerve channel (dual innervation), whereas in Pentatomomorpha, only the mandibles are innervated. The stylets arise from two stylet pouches in the head, which are located between the hypopharyngeal and maxillary plates (Snodgrass 1935). Retractor and protractor muscles are inserted at the base of each stylet; the retractor muscles all arise on the dorsal wall of the head. Mandibular protractor muscles originate from the mandibular plate (called the paraclypeus, or juga, in Heteroptera) (Snodgrass 1935) and insert on an arm that functions as a lever. The shape of the mandibular lever varies among true bugs from quadrangular to triangular (Cobben 1978). Maxillary protractor muscles arise from the maxillary plate and insert on the base of the maxilla. In some Heteroptera, a maxillary lever is also present (Cobben 1978); this lever is most complex in phytophagous species but entirely missing in many predatory heteropteran taxa (Cobben 1978). The maxillary stylets are centrally located, and slide between ridges on the hypopharynx; distal to the hypopharynx, the maxillary stylets link with one another, thereby forming the separate salivary and food canals, and the mandibular stylets converge to surround them (Snodgrass 1935).

The mandibular stylets of Heteroptera are notched, toothed, or barbed apically (Cobben 1978). For predators, this facilitates puncturing the prey and helps to stabilize and anchor the stylet bundle as the maxillae probe deeper into the prey

(Cobben 1978). In phytophagous groups, in which the mandibles travel together with the maxillae, mandibular serrations assist in initially penetrating the plant epidermis and anchoring the mandibular tips as the stylets progress but may also lacerate tissues internally. The number of mandibular teeth is variable, greater in Reduviidae and Nabidae (Cimicomorpha) than in Pentatomomorpha, even in the predatory Geocoridae and Asopinae (Pentatomidae) (Cohen 1990). A general trend of reduced mandible tooth number in phytophagous species compared with aquatic carnivores and hematophages was noted by Faucheux (1975), who examined a coreid, a pentatomid, and a mirid. He also noted that both sides of the mandibular apex were toothed in the phytophagous species and that teeth were less pointed in species that fed on the leaf blade compared with stem and petiole feeders (Faucheux 1975).

Tooth number and arrangement was consistent among several genera of phytophagous pentatomids (Depieri and Panizzi 2010a). All four pentatomid species examined were equipped with four central teeth and three pairs of lateral teeth on each mandible (Depieri and Panizzi 2010a). However, *E. fullo* (Pentatomidae) differs in having a series of transverse serrate ridges subapically, with two teeth (nodules) at the apical tip (Wang and Dai 2020). Other heteropteran families also show some variation in mandibular structure. The barbs and teeth present on the mandibular stylets of pyrrhocorids are thought to assist in penetrating hard seed coats (Rani and Madhavendra 1995; Wang and Dai 2017); mandibles of *Pyrrhocoris sibiricus* Kuschakevich have three central teeth and two pairs of lateral teeth, with parallel ridges subapically (Wang and Dai 2017). Largids have a similar arrangement but with fewer central teeth, and the pattern varies among species; greater serration is associated with a seed-feeding habit (Wang et al. 2020a). Two coreid species, feeding primarily in pericarp and seeds of green fruit, also have serrated mandibles (Rodrigues et al. 2007). In the tingid *Stephanitis nashi* Esaki and Takeya, 30 pairs of small lateral teeth were observed on the mandibular stylet tips, to facilitate stylet entry into the leaf (Wang et al. 2020b). Phloem-feeding Heteroptera also have toothed mandibles, described as "slightly serrated" in blissids (Anderson et al. 2006) and "grooved with irregular prominences" in a plataspid (Stubbins et al. 2017). Mandibles of mirids are barbed, hooked, serrated, or notched laterally at the tips (Awati 1914; Smith 1926; Cobben 1978; Roitberg et al. 2005; Wang et al. 2019).

Mandibular wear over time was shown for *Euschistus heros* (F.) (Pentatomidae) females fed a typical diet of legume seeds and pods. Significant differences in distal tooth length, second tooth height, and total toothed area of the mandible were evident after 30 and 60 days of feeding, compared with teneral adult mouthparts (Depieri and Panizzi 2010b). Similar results, demonstrating that mandibular wear was associated with plant stem feeding, were reported for the zoophytophagous mirid *Dicyphus hesperus* Knight, using a slightly different method of tooth measurement (Roitberg et al. 2005).

The maxillae are internally grooved, with ridges or processes that form three permanent linkages between the right and left stylets (Brożek and Herczek 2004). The interlocking of the maxillary stylets is tighter in Pentatomomorpha than other groups (Cobben 1978). In most phytophagous groups, except for Phloeidae and

Urostylidae, a ridge also extends outwards on each side, dovetailing with a corresponding groove in the mandibles to link the entire stylet bundle. This tighter fastening of the mandibular and maxillary stylets, and the resultant coordination of stylet movement, is considered a highly derived condition (Cobben 1978). The interlocked central maxillary ridges divide the channel into a dorsal food canal and (in most Heteroptera) a ventral salivary canal.

The food canal is always larger in diameter than the salivary canal, but this size difference is variable across Heteroptera and is minimal in some phytophagous groups (e.g., Tingidae) (Brożek and Herczek 2004). A concavity in the right maxillary stylet forms the bulk of the salivary canal in many cases, whereas the food canal is always formed more centrally from grooves in both stylets (Brożek and Herczek 2004). The salivary canal connects distally to the salivary apparatus within the hypopharynx and ultimately to the salivary glands (Snodgrass 1935). The food canal connects directly with the precibarium, eventually leading via the cibarium (sucking pump) to the esophagus (Backus 1988).

Maxillary stylets are strongly barbed in predatory species but not in phytophagous and blood-feeding groups (Cobben 1978, 1979). This evolutionary transition is evident in Cimicomorpha and especially the Miridae (Cobben 1978). The inner surface of the right maxillary stylet is serrated in the predatory Nabidae and Anthocoridae. In the Miridae, composed of zoophagous, omnivorous, and phytophagous species, this stylet exhibits a varying degree of serration, from sharply barbed in strictly zoophagous species to shallowly serrated or smooth in phytophagous species (Boyd Jr et al. 2002; Boyd Jr 2003; Wang et al. 2019). In the strictly herbivorous Tingidae, serration is also reduced; Cobben (1978) describes rudimentary tooth-like indentations in a single maxilla, whereas Wang et al. (2020b) observed five small teeth apically on the inner surface of each maxillary stylet in *S. nashi*. Small blunt external teeth are visible on the maxillae of seed-feeding pyrrhocorids (Wang and Dai 2017), but the maxillae of the closely related largids are smooth distally (Wang et al. 2020a), as are those of the phytophagous pentatomid *E. fullo* (Wang and Dai 2020).

2.2.3 Saliva

Two types of saliva are known in Heteroptera. A watery saliva is made by all true bugs (Miles 1968); this fluid contains digestive enzymes that vary with diet and mode of feeding (Hori 2000). A gelling saliva is produced by some heteropterans; this forms a solid cone or "flange" on the surface of the food item and may continue to be exuded to encase the stylets as they penetrate plant material. Miles (1968) initially considered the production of solidifying saliva to be restricted (among Heteroptera) to the Pentatomomorpha. He subsequently (1972) modified this assertion, correctly observing that only the ability to form an extended "stylet sheath" (herein referred to as a salivary sheath) by lining the stylet path with gelling saliva is unique to Pentatomomorpha among the true bugs.

Several species of Cimicomorpha do produce a salivary flange when feeding on prey (Cohen 1990) or blood (Friend and Smith 1971), including *Nabis alternatus* Parshley (Nabidae), *Zelus renardii* Kolenati, *Sinea confusa* Caudell, and *Rhodnius prolixus* Stål (Reduviidae). Cohen (1990) observed flange formation when insects fed on solid food, but not liquid diets; *R. prolixus* produced a flange when the labium contacted the rubber membrane surrounding the artificial diet, but not when touching a glass slide (Friend and Smith 1971). These salivary flanges, although somewhat resembling those produced by phytophagous Pentatomomorpha, are restricted to the surface of the prey; no gelling saliva accompanies the stylets as they penetrate prey tissue. The flanges produced by *R. prolixus* remain hollow after stylet removal as no additional saliva is exuded to plug the opening. The function of the flange in these species is presumed to be stabilization of the tip of the labium as the mandibles or the entire stylet bundle initially penetrate (Friend and Smith 1971; Cohen 1990).

Neither mirids nor tingids have been reported to produce gelling saliva (Miles 1972; Cohen 1990), but Miles' (1972) generalization that no phytophagous Cimicomorpha can do so appears to be in error based on recent findings. Solid saliva is deposited by *Thaumastocoris peregrinus* Carpintero & Dellapé (Thaumastocoridae: Thaumastocorinae) at the point of stylet entry (on the epidermis or between the stomatal cells) (Santadino et al. 2017). No sheath material accompanies the stylets as they penetrate eucalyptus leaves (Santadino et al. 2017). The monophyly of Thaumastocoridae, and the infraordinal placement and position of the Thaumastocorinae, has been questioned (Cobben 1968, 1978; Schuh et al. 2009), but the most recent cladistic analysis places both subfamilies solidly within the Miroidea (Cimicomorpha) (Weirauch et al. 2019). Further studies of the feeding behavior of other thaumastocorid species would thus be of interest.

Plant-feeding Pentatomomorpha produce both watery and gelling saliva. At the start of feeding, a conical flange is produced (Miles 1972), resembling that reported by Cohen (1990) for nabids; such structures are also formed when the bug attempts unsuccessfully to penetrate a non-food surface (Miles 1972). At this point, however, feeding behavior of some pentatomomorphans begins to resemble that of Auchenorrhyncha and Sternorrhyncha in that a true salivary sheath is formed (Miles 1968). In soft fruits, stems, and leaves, the saliva forming the surface flange continues to be exuded as the stylets pass through internal tissues, creating a tube that follows the feeding path, open at the end but encasing the stylets. The tip of the labium is embedded in the flange material (Miles 1959).

The process of sheath formation has been most closely observed and described in aphids (Nault and Gyrisco 1966; McLean and Kinsey 1965, 1967), but is similar in Pentatomomorpha (Miles 1968). Sheath material is secreted, and each drop is expanded by addition of watery saliva. The sheath material is still penetrable at this point, and the stylets push partially or completely through the blob of saliva, then pull back slightly to repeat the process. A sheath forming in a liquid medium, unconstrained by surrounding plant cells, thus has a beaded appearance. Within plant tissues, the sheath conforms to the constraints of the surrounding tissues. The salivary sheath thus produced may be straight or branched. Branching is accomplished when stylets are partially retracted and then pushed through the side of the existing sheath. Stylet probes to vascular tissue are accompanied by a salivary sheath throughout their length, presumably sealing the stylets into the target cell and enabling ingestion of vascular fluids. Depositions of sheath material may block xylem or phloem vessels (Miles 1959; Hori et al. 1984; Neal 1993). A salivary sheath may even be formed when stylets pass through relatively hard material like pecan or peanut shells, and thin cylindrical sheaths have been observed to continue beyond the internal packing through the airspace between the shell and the seed (Yates et al. 1991; Mitchell and Francis unpublished).

In mature hard seeds, the sheath is restricted to the proximal portion of the path, and feeding damage is evident beyond the termination of sheath material (Miles 1959). The resistance of the plant material affects the appearance of the sheath, because more saliva is exuded as stylets pass through denser tissues, resulting in the thicker, shorter sheath (Miles 1959). When the mouthparts are withdrawn, the opening through which the stylets passed is often sealed with additional gelling saliva, although some flanges remain open (Hollay et al. 1987).

In older literature, the flange and the complete salivary sheath are often interchangeably referred to as the "stylet sheath" (e.g., Bowling 1979), but it is important to make a distinction between them. Counts of surface flanges alone do not always reflect actual insertions (Mitchell et al. 2004) and may not be reliable measures of food consumption (Zeilinger et al. 2015). Furthermore, a surface flange is associated with several modes of feeding, used by species in both phytophagous infraorders, but full salivary sheath production is restricted to Pentatomomorpha. Conflating the two structures has led to confusion and incorrect assertions; Schuh and Weirauch (2020), for example, state that reduviids and nabids feed by means of a salivary sheath.

Salivary glands of Heteroptera are located in the thorax and divided into principal and accessory glands. The accessory gland is tubular in Pyrrhocoridae, Lygaeidae sensu lato, Berytidae, and Coreidae, but more commonly bulbous and equipped with a large central cavity (Baptist 1941). Among other functions, it secretes fluid with which to dilute the watery saliva (Miles 1967). The principal gland is divided into a variable number of lobes: two (Pentatomidae, Miridae, Tingidae), three (Lygaeidae sensu *lato*, Berytidae) four (Pyrrhocoridae, Rhopalidae), or more (Coreidae) (Baptist 1941; Taylor 1995). In the bilobed condition, the lobes are designated anterior and posterior, and their roles in enzyme production are specific.

The precursors of the gelling saliva are produced in the anterior lobe of the principal salivary glands in pentatomids (Miles 1964), in the anterior and lateral lobes of the salivary glands of lygaeids (Miles 1967), and in the anterior, anterolateral, and median lobes in a coreid (Taylor 1995). The sheath material itself is a lipoprotein solidified by hydrogen bonds and disulfide bonds that form rapidly upon secretion (Miles 1964, 1972). Disulfide bonds form immediately upon exposure to oxygen (Miles 1964), followed by hydrogen bonding. Amino acids in the watery saliva are thought to prevent hydrogen bonding, allowing the sheath material to remain temporarily flexible (Taylor and Miles 1994).

The digestive enzymes associated with the watery saliva are produced primarily in the posterior lobe of the principal salivary glands in both infraorders of phytophagous Heteroptera. The enzymatic content and composition of watery saliva varies extensively among taxonomic groups (Hori 2000; Sharma et al. 2014). Enzymes frequently found in saliva of phytophagous Heteroptera include amylase, protease, pectinase, α-glucosidase, and phenol oxidase (Hori 2000) and reflect the mode of feeding. The posterior lobe of the salivary gland in mirids produces pectinase (endo-polygalacturonase) and amylase, while the accessory gland, with its large central reservoir, contributes a dilute watery fluid along with catechol oxidase (Strong and Kruitwagen 1968; Taylor and Miles 1994; Taylor 1995; Hori 2000; Wheeler Jr 2001). Amylase is also reported in the watery saliva of a lygaeid (Miles 1959, 1967) and in the salivary glands of many heteropterans (Hori 2000 and references therein), but pectinase appears to be restricted to mirids (Hori 2000). In some cases, enzymes may be inducible; the zoophytophagous mirid *Lygus hesperus* Knight produces more of the proteolytic enzyme elastin when exposed to this protein in artificial diet (Zeng and Cohen 2001).

The posterior lobe of the salivary glands of the coreid, *Mictis profana* (F.) produces sucrase (α-glucosidase) and the accessory lobe produces catechol oxidase (Taylor and Miles 1994; Taylor 1995). Polyphenol oxidases are present in the watery saliva of all phytophagous Hemiptera (Miles 1972) and thought to be involved in detoxification of plant defensive chemicals (Taylor and Miles 1994) although they may play a role in plant damage (Miles and Taylor 1994). The production of α-glucosidase, in contrast, may be restricted to Coreidae, particularly those feeding on vegetative plant material.

The long, meandering duct of the accessory gland passes forward into the head, then returns to join the duct of the principal gland where the latter emerges from the principal gland (Baptist 1941). Ducts from the paired salivary glands terminate in the salivary pump (Snodgrass 1935), either separately or fused into a common duct. The pump is cup-shaped in most Heteroptera but more complex in Pentatomomorpha (Cobben 1978). The exit from the pump is located at the tip of the hypopharynx and aligned with the salivary canal of the maxillae as they converge. Muscles arising on the hypopharynx control the action of the salivary pump, and saliva is drawn from the salivary ducts and delivered to the salivary canal (Snodgrass 1935).

2.2.4 Foregut

The food canal in the stylet bundle leads to the precibarium [food meatus of Snodgrass (1935)], thence to the cibarium, or sucking pump, and finally to the stomodeum (pharynx plus esophagus). The true mouth is located at the juncture of the cibarium and the pharynx, placing both the precibarium and the cibarium technically in the pre-oral cavity, outside of the foregut (Snodgrass 1935). However, Goodchild (1966) considers the cibarium to be the "extreme anterior end of the alimentary canal," and Snodgrass (1935) defines the *functional* mouth as the food meatus, because it provides the entrance into the cibarial pump chamber. Similarly, both the cibarium and precibarium are considered functionally part of the foregut by Backus (2016) and will be treated as such herein.

The precibarium is a short tube formed when the posterior (epipharyngeal) wall of the anteclypeus presses against a groove in the hypopharyngeal wall (Snodgrass 1935). It connects the food canal in the stylet bundle to the cibarial pump and plays a vital role in gustation prior to ingestion. In Heteroptera, unlike Auchenorrhyncha, both the precibarium and the cibarium are tubular. Chemosensilla are located on the epipharyngeal wall of the precibarium, arranged in two clusters, anterior and posterior (Cobben 1978). These precibarial sensilla are collectively referred to in earlier literature as the epipharyngeal or gustatory organ. The anterior grouping is present in all Heteroptera, but the arrangement varies from irregular, to oval, to a double row typical of Miridae and Pentatomomorpha. The posterior cluster, when present, ranges from 2 to 8 sensilla (Cobben 1978). In Heteroptera, a small precibarial valve is reported to occur, separating the two sensilla clusters (E. A. Backus, unpublished, cited in Cervantes et al. 2016). This valve functions in Auchenorrhyncha to expose indrawn fluids sequentially to the anterior and posterior sensilla (Backus 1988) or to move fluids back and forth across both sets of chemoreceptors for tasting (Backus 2016); presumably its function in Heteroptera is similar.

During labial dabbing, watery saliva may be exuded and then drawn into the precibarium via the food canal (Miles 1958), supplementing chemosensory exploration of the plant surface by the labial tip sensilla (Backus 1988). Miles (1958) demonstrated that bugs in two families (Pyrrhocoridae and Lygaeidae) could distinguish between pure sucrose solution and a distasteful version when only the stylets (and not the labium) contacted the liquid, thereby confirming gustatory discrimination when liquids were drawn up the food canal. In addition to preliminary sampling of the plant surface in this manner, cell contents are also sampled after feeding has begun. Because heteropteran stylets have only mechanoreceptors, all gustatory examination of internal plant fluids would have to be done by the precibarial sensilla (Backus 1988). A recognizable EPG waveform, the X wave, is associated with precibarial sampling and subsequent egestion of sampled fluid in sharpshooters (Backus et al. 2009). Such X waves have also been identified in a phloem-feeding pentatomid (Lucini and Panizzi 2016), phloem-feeding blissids (Backus et al. 2013), and lygus bugs (Cervantes et al. 2016), strongly suggesting that heteropterans use the precibarial sensilla in a similar manner during feeding (Cervantes et al. 2016).

Once the precibarial sampling is complete, ingestion follows, powered by the cibarium. In Heteroptera, the position of the cibarium is horizontal, with the diaphragm dorsal, in contrast to its vertical position closer to the front of the head in cicadas and other auchenorrhynchans (Snodgrass 1935). The dilator muscles of the cibarium in Hemiptera arise on the clypeus and insert on the epipharyngeal wall of the sucking pump chamber (Snodgrass 1935). This wall, or "cibarial diaphragm" (Backus 2016), is flexible, unlike the other walls of the cibarium, which are heavily sclerotized (Snodgrass 1935). The flexible wall, normally infolded downwards into the cibarial chamber, is lifted upwards by contraction of the dilator muscles, opening the pump chamber, and thereby allowing fluid from the food canal and precibarium to fill the cibarium. Swallowing into the stomodeum then follows as the muscles are relaxed and the cibarial diaphragm collapses back into the lumen of the pump chamber (Snodgrass 1935). Simultaneous recordings of cibarial muscle activity and EPG waveforms in a xylem-feeding sharpshooter leafhopper showed a clear correspondence between diaphragm lift and a waveform pattern associated with ingestion (Dugravot et al. 2008). The pharynx and esophagus are not easily distinguished, and most writers (e.g., Goodchild 1966; Schuh and Weirauch 2020) simply use the term esophagus for the entire region of the foregut beyond the cibarium. This portion of the stomodeum is thin and tubular for most of its length. There is no true crop in Hemiptera, but in some Coreidae the posterior region of the esophagus may be expanded with the epithelial cells arranged into ridges (Goodchild 1966). The hollow, cone-shaped esophageal valve separates the foregut from the midgut. The extent to which this valve is developed varies, from well-developed in a sap-feeding coreid to poorly developed in seed-feeders (Goodchild 1966).

2.2.5 Penetration into Plant Tissue

Before the stylets emerge from the labial sheath and pierce the plant, the labium touches the substrate (sometimes repeatedly), and watery saliva may be exuded from the salivary canal and sucked back through the food canal to the gustatory organs of the epipharyngeal wall (Miles 1958). Such labial dabbing thus involves both the precibarial and labial chemosensilla in assessing the plant host (Backus 1988). In mirids, all four labial segments extend outward from the body during dabbing (Wheeler Jr 2001), but in *Nezara viridula* (L.) (Pentatomidae) only segments two through four are involved (Esquivel 2011). The labium may be slightly elbowed as the tip of the mouthparts touches the plant surface, but the labrum remains appressed to the first labial segment and the stylets are not exserted (Esquivel 2011).

As penetration into plant tissue begins, the grooved labrum steadies the stylet bundle proximally (Awati 1914; Snodgrass 1935). In Pentatomomorpha, a flange of gelling saliva is exuded to wrap and stabilize the tip of the labium (Miles 1959; Miles 1972). Stylet exsertion involves contraction of the stylet protractor muscles (Snodgrass 1935), elbowed bending of various segments of the labium (Miles 1987; Esquivel 2011), plus movement of the labrum (Esquivel 2011). Head movements (Miles 1987; Esquivel 2011) and body leverage (Esquivel 2011) are also involved in stylet penetration.

The mechanics of stylet penetration have been examined in detail in the pentatomids *Eurydema rugosa* Motschulsky, *N. viridula* (Hori 1968; Esquivel 2011) and *E. fullo* (Wang and Dai 2020), the pyrrhocorid *P. sibiricus* (Wang and Dai 2017), and the mirids *Lygocoris pabulinus* (L.) [as *Lygus pabulinus* (L.)] and *Heliopeltis clavifer* (Walker) (Awati 1914; Miles 1987), among others. In *N. viridula*, the first segment of the rostrum is immobile, held between the bucculae. Articulations between the first and second segments, and the second and third segments create the elbow folding of the labium. At the beginning of a probe, the labrum, holding the basal portion of the stylets, swings outwards from the head, pulling the stylets out from the labial groove of the first and second segments. This orients the stylets perpendicular to the substrate. Meanwhile, the second segment has angled in relation to both the first and third segments, creating an elbow bend. The result of these movements is a vertical alignment of the labrum and the third and fourth segments, with all three guiding the stylets, and the emergence of the stylets from the end of the fourth segment. As the elbow bend intensifies, decreasing the acute angle between the first and second segments, the stylets push further out from the labium and deeper into the plant (Esquivel 2011). The posture of the insect can also affect this angle and consequently stylet penetration depth (Esquivel 2011). Models developed to estimate maximal stylet penetration for pentatomids and mirids (Esquivel 2011, 2015) have demonstrated that total rostral length, traditionally used to estimate depth of penetration, is not accurate; instead, the lengths of the first and second labial segments, and the angle formed between them, are the critical determinants of stylet insertion depth (Esquivel 2015). In *E. rugosa*, the process is similar except that the first segment moves freely, such that all segments of the labium initially orient perpendicular to the substrate. The first segment then bends back against the venter and the second segment angles forward, articulating with the first and third segments. As in *N. viridula*, the vertical stylet bundle is supported by the labrum and the two final labial segments (Hori 1968). Stylet penetration follows a similar pattern when *E. fullo* feeds on young stalks, or on fruits with thin skins such as pears or grapes, but labial movements are quite different when the bug feeds on oranges. After an initial elbow bend between the first and second segments, the labium bends again at a 90 degree angle at the juncture of the second and third segments, until the two apical segments are parallel with the fruit surface, leaving the stylets vertically unsupported except by the labrum, basally (Wang and Dai 2020).

During seed-feeding by *P. sibiricus*, the typical elbow fold is followed by complete release of the stylet bundle from both the labium, which bends back along the venter, and the labrum. The head approaches closely to the food surface, permitting deep stylet penetration, with little of the stylet bundle remaining visible outside the seed. However, when feeding on stems, as in *E. fullo*, the labium simply forms a typical elbow bend and the stylets remain supported by the labrum and the apical segments of the labium (Wang and Dai 2017).

In mirids, the first segment of the labium articulates with the head (Awati 1914; Miles 1987), and all four segments are initially aligned vertically with one another (Miles 1987). In *L. pabulinus*, the first segment is thickened, but the labial groove in this segment is extremely shallow and the segment bends slightly inward, leaving the stylet bundle unsupported. The grooved labrum, which presses against the first labial segment, actually holds the stylet bundle at its proximal end (Awati 1914). Stylet protrusion is accomplished as the stylet protractor muscles contract and the first and second labial segments begin to move. The first segment shifts back towards the body. The labrum, holding the stylets, remains perpendicular to the body, and the stylet bundle pulls out from the labial groove of the second segment. The joint between the first and second segments forms an elbow bend (Miles 1987); essentially this joint acts as a hinge (Awati 1914), with the bend increasing to enable deeper penetration. As in pentatomids, the second segment also articulates with the third segment, allowing the latter to remain oriented perpendicular to the body. Meanwhile, the labrum continues to hold the proximal portion of the stylet bundle, which is also supported by the third and fourth labial segments as penetration into plant tissue proceeds (Miles 1987). The labial groove in the fourth segment is so deep as to completely enclose the stylet bundle (Awati 1914). These authors note no further elbowing, but Kullenberg (1944, cited in Wheeler Jr 2001) diagrammed an additional bend between the third and fourth labial segments. Cline and Backus (2002) similarly noted double bends of the labium in *L. hesperus* during drilling-laceration behavior, to maximize the extension of the stylets.

2.3 Strategies of Feeding

To fully characterize the feeding behavior of a piercing-sucking insect, one must consider four aspects. Host plant preference is the traditional descriptor, encompassing both plant species and degree of monophagy or polyphagy and ideally distinguishing breeding hosts from those on which only incidental feeding is observed (Smaniotto and Panizzi 2015). The plant parts (e.g., bud, stem, root) on which the insect preferentially feeds must also be identified; this may change seasonally or from one plant species to another (Panizzi 2000). The preferred target tissue within that plant structure represents yet another area of specialization; for example, stem-feeding pentatomids may be ingesting from phloem, xylem, or parenchyma (Lucini and Panizzi 2018a; Almeida et al. 2020). Finally, the feeding strategy, or mode of feeding, explains how the insect acquires nutrients from the selected target tissue; this aspect will be the focus of the subsequent sections. Electropenetrography, in combination with histological examination of tissue damage, stylet placement, and salivary deposits, is ideal for identifying both target tissue and feeding strategy.

Our understanding of the feeding strategies of piercing-sucking insects has expanded and altered over the past half century. An historical overview will be presented here, followed by more detailed description of known feeding modes relevant to Heteroptera.

2.3.1 *Historical Overview*

Miles' (1972) original concept assumed a plant-feeding heteropteran ancestor and proposed three plant-feeding modes, distinguished by the types of saliva involved, the movement of the stylets, and the source of nutrition. Scratch-and-suck feeding, emptying cells close to the plant surface, was associated with thrips and tingids, as well as with a putative hemipteran ancestor. Watery saliva is deposited in target cells, and Miles suggested that cells are fed upon individually or without extensive laceration. Cobben (1978) renamed this feeding mode more aptly as "punch-and-suck"; here we suggest the alternative term "puncture-and-suck." Two more common heteropteran feeding modes: "stylet-sheath" (herein termed salivary sheath) feeding and lacerate-and-flush feeding were also distinguished by Miles (1972). Salivary sheath feeding entails production of both watery and gelling saliva and is limited to those Hemiptera who can discharge gelling saliva *within* plant tissue as well as superficially. A tube of solidifying saliva lines the path of the stylets completely, from the surface flange to the target vascular tissue (phloem or xylem) from which the insect ingests. In lacerate-and-flush feeding, plant tissues are lacerated by stylet movements, while enzymes in watery saliva further contribute to cell breakdown. The mix of watery saliva and cell contents is then ingested. Gelling saliva is not required, although a small deposit of gelling saliva may anchor or stabilize the stylets at the plant surface or even accompany the stylets partway to the feeding site. Parenchyma and plant reproductive structures are typically targeted by lacerate-and-flush feeding.

Two additional feeding strategies were subsequently added to Miles' scheme: macerate-and-flush and osmotic pump feeding (Miles and Taylor 1994), following the discovery that both mirids and coreids could empty cells considerably beyond the reach of their stylets (Miles 1987). Macerate-and-flush feeding resembles lacerate-and-flush, but the stylets do not move extensively. Instead, tissue breakdown results from enzymatic activity, primarily pectinase, which degrades the intercellular matrix (Taylor 1995). This feeding mode is associated with mirids. The watery saliva permeates the plant tissue far beyond the stylet tips, thereby allowing small insects to create relatively large lesions (Miles 1987). Osmotic pump feeding has been associated with four tribes in the family Coreidae, all of which produce a characteristic enzyme, α-glucosidase, in the watery saliva. This salivary sucrase is thought to mimic the action of the plant's invertase (Taylor 1995); the breakdown of sucrose to fructose and glucose alters the osmotic balance of fluids in the intercellular spaces, prompting phloem unloading and leakage of phloem parenchyma cells. The leaked cell contents are then ingested by the insect. Due to infiltration of saliva through the tissues, and accumulation of leaked nutrients in the intercellular spaces, the damaged area may extend some distance away from the stylet insertion point (Miles and Taylor 1994).

Hori (2000) listed four feeding modes for Heteroptera: salivary sheath (as stylet sheath), lacerate-and-flush, macerate-and-flush, and osmotic pump, associating the latter with coreids in general and macerate-and-flush with Cimicomorpha (Miridae). Neither Tingidae nor scratch-and-suck feeding were mentioned.

Focusing primarily on Auchenorrhyncha and the causes of hopperburn, Backus et al. (2005) introduced the idea of "feeding tactics" as subcategories of feeding strategies. Two basic strategies were proposed, the salivary sheath strategy and the newly defined cell rupture strategy, into which lacerate-and-flush and subsequently macerate-and-flush (Backus et al. 2007) were subsumed as tactics. Puncturing and draining of individual cells was categorized as part of lacerate-and-flush. A more recent compilation of feeding strategies (Sharma et al. 2014) includes osmotic pump feeding as a third strategy. Two different types of "sap suckers" are distinguished by Sharma et al. (2014): those that produce a stylet sheath and those that do not. The former category includes traditional salivary sheath and osmotic pump feeders, whereas the latter category refers to Thysanoptera ingesting the contents of palisade parenchyma cells.

2.3.2 *Salivary Sheath Strategy*

Sternorrhyncha, Auchenorrhyncha, and Pentatomomorpha all produce a gelling saliva. Most of these insects are able to form a complete salivary sheath with this material, encasing the stylets from the plant surface to the ingestion point. The sheath material anchors the labium to the substrate at one end and seals the stylets into the target cell, either a phloem sieve element or a xylem vessel. This strategy is thought to allow efficient ingestion from vascular tissues under pressure (Miles 1972; Cobben 1978) and is associated with insects that feed or hydrate from the plant vascular system. For details of sheath production, see Sect. 2.2.3.

Characterizing specific taxa of Auchenorrhyncha and Sternorrhyncha as salivary sheath feeders is straightforward, but categorizing is more difficult for Pentatomomorpha. These insects may, for example, produce a partial salivary sheath or only a superficial salivary flange when feeding on seeds, but a complete sheath when hydrating from xylem vessels or ingesting from phloem sieve elements. The latter behaviors fall clearly under the salivary sheath strategy, but the former, despite the presence of gelling saliva, represents a different strategy, cell rupture feeding (see Sect. 2.3.3). This plasticity of feeding behavior has been known for some time in a lygaeid (Miles 1959) and a pentatomid (Hori et al. 1984), but more recent EPG studies have shown such behavioral switching to be unexpectedly common (Lucini et al. 2016; Lucini and Panizzi 2017a,b, 2018b; Mitchell et al. 2018b; Serteyn et al. 2020).

2.3.3 *Cell Rupture Strategy*

The cell rupture feeding strategy as defined by Backus et al. (2005, 2007) includes two tactics relevant to Heteroptera, lacerate-and-flush and macerate-and-flush. We add here a third tactic, the original scratch-and-suck strategy proposed by Miles (1972) for some hemipteroids (tingids and thrips), which we have renamed puncture-and-suck to reflect the feeding behavior of these insects more correctly.

2.3.3.1 Lacerate-and-Flush Tactic

Originally associated with Miridae, Lygaeidae, Pyrrhocoridae, and Pentatomidae (Miles 1972), this tactic is now known to be used by several pentatomomorphan taxa, but not phytophagous Cimicomorpha. Rapid stylet movements physically damage tissues, after which watery saliva is secreted to flush out the damaged cell contents. The mechanical action, coupled with amylase, proteases, and other enzymes in the watery saliva, creates a mixture of saliva and nutrients that is then ingested. In this way, even the contents of mature seeds may be reduced to an ingestible liquid. Several EPG studies have demonstrated lacerate-and-flush patterns in pentatomids feeding on developing seeds and fruits (Lucini and Panizzi 2018a, b); although the behavior is described in some species as lacerate-and-macerate, a hybrid tactic which involves more enzymatic breakdown (Lucini and Panizzi 2018a).

2.3.3.2 Macerate-and-Flush Tactic

Mirid feeding is now classified as macerate-and-flush (Miles and Taylor 1994), to emphasize the greater involvement of enzymatic digestion and specifically the role of salivary pectinase in weakening tissues and degrading cell walls. Stylet movements are slow (Cervantes et al. 2017) and the head of the feeding mirid moves little during feeding (Miles 1987), unlike the up-and-down movements reported for some pentatomids (Esquivel 2011; Lucini et al. 2016). Lesions produced by macerate-and-flush feeding may extend substantially beyond the maximal reach of the stylets, as watery saliva percolates through the plant tissue. Damaged regions show empty or shrunken cells initially and the tissue deteriorates further over time (Miles 1987). Pectinase is reported from the salivary glands or saliva of a wide variety of mirids but from few other heteropteran taxa (Hori 2000) and appears to be intimately associated with the mode of feeding of Miridae.

2.3.3.3 Puncture-and-Suck Tactic

Individual cell penetration, initially described by Miles (1972) as scratch-and-suck feeding, has been ignored in most subsequent treatments of feeding strategies (e.g., Hori 2000), yet represents a valid description of feeding behavior in some hemipteroids, including tingids, thaumastocorines, typhlocybine leafhoppers, and thrips. *Empoasca* spp. (Typhlocybinae) exhibit a behavior originally termed single-cell puncture feeding that was associated with a specific waveform (I_b) (Hunter and Backus 1989); however, this waveform was subsequently redescribed as flushing behavior and subsumed under lacerate-and-flush (Backus et al. 2005).

Here we reclassify individual cell penetration as a separate tactic of the cell rupture strategy. The formal definition of a feeding tactic (Backus et al. 2005) is "stereotypical sequences of probing behaviors (best represented by EPG waveforms) within a feeding strategy." Feeding behavior of tingids and thaumastocorids has not been analyzed using EPG, but studies of plant damage indicate that these insects selectively pierce cells of the palisade parenchyma and remove cell contents, including chloroplasts, leaving spongy mesophyll relatively intact.

Histological studies using light microscopy showed palisade parenchyma cells of azalea leaves completely emptied by tingid feeding (Buntin et al. 1996). Similarly, thaumastocorines removed chloroplasts from chlorenchyma cells of eucalyptus leaves puncturing of multiple adjacent cells produced visible chlorotic spotting on the leaves (Santadino et al. 2017). Histological preparations of tingid feeding show stylets terminating in individual palisade cells (Pollard 1959), although this author also notes some laceration of cell walls and diffusion of a salivary secretion. Only amylase has been reported from tingid saliva, not pectinase (Hori 2000); in this also they differ from the closely related mirids. Puncture-and-suck feeding, although likely restricted to a few families of true bugs, is also characteristic of Thysanoptera. EPG research (Stafford et al. 2011) has shown that thrips feed by emptying individual cells, although their mouthpart structure is quite different from that of Heteroptera. EPG studies of tingids or thaumastocorids would be of interest.

2.3.4 Osmotic Pump Strategy

The salivary enzyme α-glucosidase, found in high concentration in the watery saliva of several coreid species, is uniquely associated with this feeding strategy. *M. profana* (Mictini) has been most extensively studied (Miles and Taylor 1994; Taylor 1995) but four other genera in three tribes (Amorbini, Coreini, Dasynini) share this characteristic salivary makeup. During feeding on shoots and new growth, a salivary sheath accompanies the stylets as they penetrate toward parenchyma in the vicinity of a vascular bundle. Watery saliva is released and infiltrates the tissue for some distance beyond the stylet termination point, creating a "water-soaked lesion" as well as eventual wilting beyond the feeding puncture towards the tip of the shoot (Miles 1987). The conversion of sucrose to glucose and fructose by α-glucosidase creates an osmotic gradient in the intercellular spaces, inducing the phloem to unload and surrounding parenchyma cells to release their contents and collapse. This creates expanded, fluid-filled intercellular spaces filled with nutrients which the insect ingests – essentially a nutrient sink (Taylor 1995). The subsequent wilting of the shoot tips may be caused by necrosis of xylem parenchyma (Taylor and Miles 1994).

Hori (2000) considers production of salivary α-glucosidase, and thus osmotic pump feeding, to be characteristic of the Coreidae. However, only members of four tribes have been definitively associated with osmotic pump feeding. It is likely that other coreids feeding on stems and new growth share this strategy (e.g., Acanthocerini, Acanthocephalini, Nematopodini). However, it is important to point out that not all coreids feed on shoots. Anisoscelini (e.g., *Narnia, Leptoglossus, Anisoscelis, Holymenia*) feed on fruits and developing seeds (Mitchell 2006; Rodrigues et al. 2007; Miller et al. 2016) but also employ the salivary sheath strategy to access vascular tissue. *Chelinidea* spp. (Chelinideini) create visible circular lesions in the mesophyll of pads of *Opuntia* cactus (Mann 1969), likely the result of cell-rupture feeding. *Clavigralla* spp. (Clavigrallini) are notorious for direct damage to pods and seeds of legumes, also likely via cell rupture feeding. Members of two tribes (Cloresmini and Colpurini) produce honeydew and are tended by ants, and thus thought to feed directly from phloem sieve elements (Maschwitz et al. 1987), presumably via the salivary sheath strategy. The diversity of feeding habits in this family requires that taxa be examined at the tribal level, rather than generalizing for the family as a whole.

2.4 Concluding Remarks

The mouthparts of phytophagous Heteroptera have evolved to permit a variety of feeding modes, from delicate insertion into single cells to extensive physical and enzymatic tissue destruction. Variations in the composition of watery saliva, production and employment of gelling saliva, rostral and stylet lengths, maxillary serration, and other characters allow heteropterans to exploit effectively nearly all plant tissues. Knowledge of mouthpart morphology, coupled with EPG studies of feeding behavior, will enable researchers to understand the feeding strategies and tactics of pest insects and to predict and ideally prevent direct and indirect damage to crops.

References

Almeida ACS, Jesus FG, Barrigossi JAF (2020) Unveiling the feeding behavior of *Tibraca limbativentris* (Hemiptera: Pentatomidae) on rice using an electropenetrography waveform library. J Insect Sci 20:14;1–8

Anderson WG, Heng-Moss TM, Baxendale FP, Baird LM, Sarath G, Higley L (2006) Chinch bug (Hemiptera: Blissidae) mouthpart morphology, probing frequencies, and locations on resistant and susceptible germplasm. J Econ Entomol 99:212–221

Awati PR (1914) The mechanism of suction in the potato capsid bug, *Lygus pabulinus* Linn. Proc Zool Soc London 84:685–733

Backus EA (1988) Sensory systems and behaviours which mediate hemipteran plant-feeding: a taxonomic overview. J Insect Physiol 34:151–165

Backus EA (2016) Sharpshooter feeding behavior in relation to transmission of *Xylella fastidiosa*: a model for foregut-borne transmission mechanisms. In: Brown J (ed) Vector-mediated transmission of plant pathogens. APS Press, St. Paul, pp 175–193

Backus EA, Serrano MS, Ranger CM (2005) Mechanisms of hopperburn: an overview of insect taxonomy, behavior, and physiology. Annu Rev Entomol 50:125–151

Backus EA, Cline AR, Ellerseick MR, Serrano MS (2007) *Lygus hesperus* (Hemiptera: Miridae) feeding on cotton: new methods and parameters for analysis of nonsequential electrical penetration graph data. Ann Entomol Soc Am 100:296–310

Backus EA, Holmes WJ, Schreiber F, Reardon BJ, Walker GP (2009) Sharpshooter X wave: correlation of an electrical penetration graph waveform with xylem penetration supports a hypothesized mechanism for *Xylella fastidiosa* inoculation. Ann Entomol Soc Am 102:847–867

Backus EA, Rangasamy M, Stamm M, McAuslane HJ, Cherry R (2013) Waveform library for chinch bugs (Hemiptera: Heteroptera: Blissidae): characterization of electrical penetration graph waveforms at multiple input impedances. Ann Entomol Soc Am 106:524–539

Baptist BA (1941) The morphology and physiology of the salivary glands of Hemiptera-Heteroptera. Quart J Micr Sci 82:91–139

Bowling CC (1979) The stylet sheath as an indicator of feeding activity of the rice stink bug. J Econ Entomol 72:259–260

Boyd DW Jr (2003) Digestive enzymes and stylet morphology of *Deraeocoris nigritulus* (Uhler) (Hemiptera: Miridae) reflects adaptations for predatory habits. Ann Entomol Soc Am 96:667–671

Boyd DW Jr, Cohen AC, Alverson DR (2002) Digestive enzymes and stylet morphology of *Deraeocoris nebulosus* (Uhler) (Hemiptera: Miridae), a predacious plant bug. Ann Entomol Soc Am 95:395–401

Brożek J, Herczek A (2004) Internal structure of the mouthparts of true bugs (Hemiptera, Heteroptera). Pol J Entomol 73:79–106

Buntin GD, Braman SK, Gilbertz DA, Phillips DV (1996) Chlorosis, photosynthesis, and transpiration of azalea leaves after azalea lace bug (Heteroptera: Tingidae) feeding injury. J Econ Entomol 89:990–995

Carroll SP, Boyd C (1992) Host race radiation in the soapberry bug: natural history, with the history. Evolution 46:1052–1069

Carroll SP, Loye JE, Dingle H, Mathieson M, Famula TR, Zalucki MP (2005) And the beak shall inherit – evolution in response to invasion. Ecol Lett 8:944–951

Cervantes FA, Backus EA, Godfrey L, Akbar W, Clark TL (2016) Characterization of an EPG waveform library for adult *Lygus lineolaris* and *Lygus hesperus* (Hemiptera: Miridae) feeding on cotton squares. Ann Entomol Soc Am 109:684–697

Cervantes FA, Backus EA, Godfrey L, Wallis C, Akbar W, Clark TL, Rojas MG (2017) Correlation of electropenetrography waveforms from *Lygus lineolaris* (Hemiptera: Miridae) feeding on cotton squares with chemical evidence of inducible tannins. J Econ Entomol 110:2068–2075

Cline AR, Backus EA (2002) Correlations among AC electronic monitoring waveforms, body postures, and stylet penetration behaviors of *Lygus hesperus* (Hemiptera: Miridae). Environ Entomol 31:538–549

Cobben RH (1968) Evolutionary trends in Heteroptera Part I. Eggs, architecture of the shell, gross embryology and eclosion. Agric Res Reports, Wageningen. 707:1–475

Cobben RH (1978) Evolutionary trends in Heteroptera Part II. Mouthpart structures and feeding strategies. Mededelingen Landbouwhogeschool 78-5:1–407

Cobben RH (1979) On the original feeding habits of the Hemiptera (Insecta): a reply to Merrill Sweet. Ann Entomol Soc Am 72:711–715

Cohen AC (1990) Feeding adaptations of some predaceous Hemiptera. Ann Entomol Soc Am 83:1215–1223

Dugravot S, Backus EA, Reardon BJ, Miller TA (2008) Correlations of cibarial muscle activities of *Homalodisca* spp. sharpshooters (Hemiptera: Cicadellidae) with EPG ingestion waveform and excretion. J Insect Physiol 54:1467–1478

Depieri RA, Panizzi AR (2010a) Rostrum length, mandible serration, and food and salivary canals areas of selected species of stink bugs (Heteroptera, Pentatomidae). Rev Bras Entomol 54:584–587

Depieri RA, Panizzi AR (2010b) Aging and food source effects on mandibular stylets teeth wear of phytophagous stink bug (Heteroptera: Pentatomidae). Neotrop Entomol 39:952–956

Faucheux MM-J (1975) Relations entre l'ultrastructure des stylets mandibulaires et maxillaires et la prise de nourriture chez les Insectes Hémiptères. C R Acad Sci Paris Série D 281:41–44

Esquivel JF (2011) Estimating potential stylet penetration of southern green stink bug – a mathematical approach. Entomol Exp Appl 140:163–170

Esquivel JF (2015) Stylet penetration estimates for a suite of phytophagous hemipteran pests of row crops. Environ Entomol 44:619–636

Friend WG, Smith JJB (1971) Feeding in *Rhodnius prolixus*: mouthpart activity and salivation, and their correlation with changes of electrical resistance. J Insect Physiol 17:233–243

Goodchild AJP (1966) Evolution of the alimentary canal in the Hemiptera. Biol Rev 41:97–140

Hollay ME, Smith CM, Robinson JF (1987) Structure and formation of feeding sheaths of rice stink bug (Heteroptera: Pentatomidae) on rice grains and their association with fungi. Ann Entomol Soc Am 80:212–216

Hori K (1968) Feeding behavior of the cabbage bug, *Eurydema rugosa* Motschulsky (Hemiptera: Pentatomidae) on the cruciferous plants. Appl Entomol Zool 3:26–36

Hori K (2000) Possible causes of disease symptoms resulting from the feeding of phytophagous Heteroptera. In: Schaefer CW, Panizzi AR (eds) Heteroptera of economic importance. CRC Press, Boca Raton, pp 11–36

Hori K, Kondo Y, Kuramochi K (1984) Feeding site of *Palomena angulosa* Motschulsky (Hemiptera: Pentatomidae) on potato plants and injury caused by the feeding. Appl Entomol Zool 19:476–482

Hunter WB, Backus EA (1989) Mesophyll-feeding by the potato leafhopper, *Empoasca fabae* (Homoptera: Cicadellidae): results from electronic monitoring and thin-layer chromatography. Environ Entomol 18:465–472

Kullenberg B (1944) Studien über die Biologie der Capsiden. Zool Bidr Upps 23:1–522

Lucini T, Panizzi AR (2016) Waveform characterization of the soybean stem feeder *Edessa meditabunda* (F.) (Hemiptera: Heteroptera: Pentatomidae): overcoming the challenge of wiring pentatomids for EPG. Entomol Exp Appl 158:118–132

Lucini T, Panizzi AR (2017a) Probing behavior of *Dichelops furcatus* (F.) (Heteroptera: Pentatomidae) on wheat plants characterized by electropenetrography (EPG) and histological studies. J Insect Sci 17:65;1–15

Lucini T, Panizzi AR (2017b) Feeding behavior of the stink bug *Dichelops melacanthus* (Heteroptera: Pentatomidae) on maize seedlings: an EPG analysis at multiple input impedances and histology correlation. Ann Entomol Soc Am 110:160–171

Lucini T, Panizzi AR (2018a) Electropenetrography (EPG): a breakthrough tool unveiling stink bug (Pentatomidae) feeding on plants. Neotrop Entomol 47:1–13

Lucini T, Panizzi AR (2018b) Electropenetrography monitoring of the neotropical brown-stink bug (Hemiptera: Pentatomidae) on soybean pods: an electrical penetration graph-histology analysis. J Insect Sci 18:5;1–14

Lucini T, Panizzi AR, Backus EA (2016) Characterization of an EPG waveform library for redbanded stink bug, *Piezodorus guildinii* (Hemiptera: Pentatomidae), on soybean plants. Ann Entomol Soc Am 109:198–210

Mann J (1969) Cactus feeding insects and mites. US Natl Mus Bull No 256:1–158

Maschwitz U, Fiala B, Dolling WR (1987) New trophobiotic symbioses of ants with South East Asian bugs. J Nat Hist 21:1097–1107

McLean DL, Kinsey MG (1965) Identification of electrically recorded curve patterns associated with aphid salivation and ingestion. Nature 205:1130–1131

McLean DL, Kinsey MG (1967) Probing behavior of the pea aphid, *Acyrthosiphon pisum*. I. Definitive correlation of electronically recorded waveforms with aphid probing activities. Ann Entomol Soc Am 60:400–406

Miles PW (1958) Contact chemoreception in some Heteroptera, including chemoreception internal to the stylet food canal. J Insect Physiol 2:338–347

Miles PW (1959) Salivary secretions of a plant-sucking bug, *Oncopeltus fasciatus* (Dall.) (Heteroptera: Lygaeidae). I. The types of secretion and their role during feeding. J Insect Physiol 3:243–255

Miles PW (1964) Studies on the salivary physiology of plant-bugs: the chemistry of formation of the sheath material. J Insect Physiol 10:147–160

Miles PW (1967) The physiological division of labour in the salivary glands of *Oncopeltus fasciatus* (Dall.) (Heteroptera: Lygaeidae). Aust J Biol Sci 20:785–797

Miles PW (1968) Insect secretions in plants. Annu Rev Phytopathol 6:137–164

Miles PW (1972) The saliva of Hemiptera. Adv Insect Physiol 9:183–255

Miles PW (1987) Plant-sucking bugs can remove the contents of cells without mechanical damage. Experientia 43:937–939

Miles PW, Taylor GS (1994) 'Osmotic pump' feeding by coreids. Entomol Exp Appl 73:163–173

Miller CW, McDonald GC, Moore AJ (2016) The tale of the shrinking weapon: seasonal changes in nutrition affect weapon size and sexual dimorphism, but not contemporary evolution. J Evol Biol 29:2266–2275

Mitchell PL (2006) Polyphagy in true bugs: a case study of *Leptoglossus phyllopus* (L.) (Hemiptera, Heteroptera, Coreidae). Denisia 19. Neue Serie 50:1117–1134

Mitchell PL, Gupta R, Singh AK, Kumar P (2004) Behavioral and developmental effects of neem extracts on *Clavigralla scutellaris* (Hemiptera: Heteroptera: Coreidae) and its egg parasitoid, *Gryon fulviventre* (Hymenoptera: Scelionidae). J Econ Entomol 97:916–923

Mitchell PL, Zeilinger AR, Medrano EG, Esquivel JF (2018a) Pentatomoids as vectors of plant pathogens. In: McPherson JE (ed) Invasive stink bugs and related species (Pentatomoidea): biology, higher systematics, semiochemistry, and management. CRC Press, Boca Raton, pp 611–640

Mitchell PL, Cooke SB, Smaniotto LF (2018b) Probing behavior of *Nezara viridula* on soybean: characterization and comparison of electrical penetration graph (EPG) waveforms on vegetative and reproductive plant structures. J Agric Urban Entomol 34:19–43

Nault LR, Gyrisco GG (1966) Relation of the feeding process of the pea aphid to the inoculation of pea enation mosaic virus. Ann Entomol Soc Am 59:1185–1197

Neal JJ (1993) Xylem transport interruption by *Anasa tristis* feeding causes *Cucurbita pepo* to wilt. Entomol Exp Appl 69:195–200

Panizzi AR (2000) Suboptimal nutrition and feeding behavior of hemipterans on less preferred plant food sources. An Soc Entomol Bras 29:1–12

Pollard DG (1959) Feeding habits of the lace-bug *Urentius aegyptiacus* Bergevin (Hemiptera: Tingidae). Ann Appl Biol 47:778–782

Rani PU, Madhavendra SS (1995) Morphology and distribution of antennal sense organs and diversity of mouthpart structures in *Odontopus nigricornis* (Stall) and *Nezara viridula* L. (Hemiptera). Int J Insect Morphol Embryol 24:119–132

Rider DA, Schwertner CF, Vilímová J, Rédei D, Kment P, Thomas DB (2018) Higher systematics of the Pentatomoidea. In: McPherson JE (ed) Invasive stink bugs and related species (Pentatomoidea). CRC Press, Boca Raton, pp 25–201

Rodrigues D, Sampaio DS, Isaias RMDS, Moreira GRP (2007) Xylem and seed feeding by two passion vine leaffooted bugs, *Holymenia clavigera* and *Anisoscelis foliacea marginella* (Hemiptera: Coreidae: Anisoscelini) with notes on mouthpart morphology. Ann Entomol Soc Am 100:907–913

Roitberg BD, Gillespie DR, Quiring DMJ, Alma CR, Jenner WH, Perry J, Peterson JH, Salomon M, VanLaerhoven S (2005) The cost of being an omnivore: mandible wear from plant feeding in a true bug. Naturwissenschaften 92:431–434

Santadino MV, Brentassi ME, Fanello DD, Coviella C (2017) First evidence that *Thaumastocoris peregrinus* (Heteroptera: Thaumastocoridae) feeds from mesophyll of Eucalyptu*s* leaves. Environ Entomol 46:251–257

Schuh RT, Weirauch C (2020) True bugs of the world (Hemiptera: Heteroptera): classification and natural history, 2nd edn. Siri Scientific Press, Manchester

Schuh RT, Weirauch C, Wheeler W (2009) Phylogenetic relationships within the Cimicomorpha (Hemiptera: Heteroptera): a total evidence analysis. Syst Entomol 34:15–48

Serteyn L, Ponnet L, Backus EA, Francis F (2020) Characterization of electropenetrography waveforms for the invasive heteropteran pest, *Halyomorpha halys*, on *Vicia faba* leaves. Arthropod Plant Interact 14:113–126

Sharma A, Khan AN, Subrahmanyam S, Raman A, Taylor GS, Fletcher MJ (2014) Salivary proteins of plant-feeding hemipteroids – implication in phytophagy. Bull Entomol Res 104:117–136

Smaniotto LF, Panizzi AR (2015) Interactions of selected species of stink bugs (Hemiptera: Heteroptera: Pentatomidae) from leguminous crops with plants in the Neotropics. Fla Entomol 98:7–17

Smith KM (1926) A comparative study of the feeding methods of certain Hemiptera and of the resulting effects upon the plant tissue, with special reference to the potato plant. Ann Appl Biol 13:109–139

Snodgrass RE (1935) Principles of insect morphology. McGraw-Hill Book Co., New York

Stafford CA, Walker GP, Ullman DE (2011) Infection with a plant virus modifies vector feeding behavior. Proc Natl Acad Sci 108:9350–9355

Strong FE, Kruitwagen EC (1968) Polygalacturonase in the salivary apparatus of *Lygus hesperus* (Hemiptera). J Insect Physiol 14:1113–1119

Stubbins FL, Mitchell PL, Turnbull MW, Reay-Jones FPF, Greene JK (2017) Mouthpart morphology and feeding behavior of the invasive kudzu bug, *Megacopta cribraria* (Hemiptera: Plataspidae). Invertebr Zool 136:309–320

Taylor GS (1995) A comparison between the salivary physiology of the crusader bug, *Mictis profana* Fabricius (Coreidae) and the green lucerne mirid, *Creontiades dilutus* (Stål). Ph.D. dissertation, Univ of Adelaide, Waite, South Australia

Taylor GS, Miles PW (1994) Composition and variability of the saliva of coreids in relation to phytoxicoses and other aspects of the salivary physiology of phytophagous Heteroptera. Entomol Exp Appl 73:265–277

Wang Y, Dai W (2017) Fine structure of mouthparts and feeding performance of *Pyrrhocoris sibiricus* Kuschakevich with remarks on the specialization of sensilla and stylets for seed feeding. PLoS One 12:e0177209;1–23

Wang Y, Dai W (2020) How does the intricate mouthpart apparatus coordinate for feeding in the hemimetabolous insect pest *Erthesina fullo*? Insects 11:503;1–22

Wang Y, Li L, Dai W (2019) Fine morphology of the mouthparts in *Cheilocapsus nigrescens* (Hemiptera: Heteroptera: Miridae) reflects adaptation for phytophagous habits. Insects 10:143;1–17

Wang Y, Brożek J, Dai W (2020a) Morphological disparity of the mouthparts in polyphagous species of Largidae (Heteroptera: Pentatomomorpha: Pyrrhocoroidea) reveals feeding specialization. Insects 11:145;1–20

Wang Y, Brożek J, Dai W (2020b) Sensory armature and stylets of the mouthparts of *Stephanitis nashi* (Hemiptera: Cimicomorpha: Tingidae), their morphology and function. Micron 132:102840. https://doi.org/10.1016/j.micron.2020.102840

Weirauch C, Schuh RT, Cassis G, Wheeler WC (2019) Revisiting habitat and lifestyle transitions in Heteroptera (Insecta: Hemiptera): insights from a combined morphological and molecular phylogeny. Cladistics 35:67–105

Wheeler AG Jr (2001) Biology of the plant bugs (Hemiptera: Miridae). Cornell University Press, Ithaca

Yates IE, Tedders WL, Sparks D (1991) Diagnostic evidence of damage on pecan shells by stink bugs and coreid bugs. J Am Soc Hortic Sci 116:42–46

Zeilinger AR, Olson DM, Raygoza T, Andow DA (2015) Do counts of salivary sheath flanges predict food consumption in herbivorous stink bugs (Hemiptera: Pentatomidae)? Ann Entomol Soc Am 108:109–116

Zeng F, Cohen AC (2001) Induction of elastase in a zoophytophagous heteropteran, *Lygus hesperus* (Hemiptera: Miridae). Ann Entomol Soc Am 94:146–151

Chapter 3
Feeding Sites of True Bugs and Resulting Damage to Plants

Contents

Abstract Phytophagous true bugs feed on virtually all plant parts, both vegetative and reproductive, from roots to shoots and buds to seeds. Preferred ingestion sites include vascular tissue; parenchyma of stems, leaves, and fruits; and seed endosperm. Stylet length and depth of penetration, production of a salivary sheath from gelling saliva, and enzymatic content of watery saliva directly influence the choice of ingestion site and the type and extent of the resulting damage; all of these factors are linked to feeding strategy. In this chapter, we provide detailed descriptions, with examples, of ingestion by heteropterans from phloem sieve elements, xylem vessels, parenchyma, seed endosperm, and staminal tissue. Damage to plant structures is reviewed in the context of feeding strategy, examining the effects of salivary sheath feeding, osmotic pump feeding, and cell rupture feeding. The ability of heteropterans to transmit plant pathogens also depends on the feeding strategy and target ingestion tissue. Acquisition and inoculation of phytoplasmas and disease-causing trypanosomes is linked with ingestion from phloem tissues. Fungal and bacterial pathogens, especially the yeast *Eremothecium coryli,* are transmitted to the reproductive structures of many crop plants, including coffee, cotton, rice, legumes, tomato, citrus, and pistachio.

A. R. Panizzi et al., *Electronic Monitoring of Feeding Behavior of Phytophagous True Bugs (Heteroptera)*, Entomology in Focus 6,
https://doi.org/10.1007/978-3-030-64674-5_3

Keywords Feeding damage · Salivary enzymes · Vascular system · Parenchyma · Endosperm · Pathogen transmission

3.1 Introduction

True bugs vary extensively in stylet length and depth of penetration, saliva production, and salivary enzymes and consequently in preferred target tissue and resulting plant damage. A major difference in the nature of the saliva exists between the two phytophagous infraorders; watery saliva is used by both Cimicomorpha (tingids, mirids, and thaumastocorids) and Pentatomomorpha for enzymatic digestion, whereas pentatomomorphans may also line the path of the stylets with a gelling saliva, as in Sternorrhyncha and Auchenorrhyncha (Miles 1959, 1972). This adaptation facilitates stylet insertion into vascular tissues, enabling feeding strategies other than cell rupture (Backus et al. 2005). Damage potential varies with enzyme content of the watery saliva (Hori 2000), which in turn is a function of feeding strategy and tactics. This chapter will examine ingestion sites and damage in the context of heteropteran feeding strategies, previously detailed in Chap. 2.

3.2 Feeding Sites of True Bugs on Plants

Feeding locations on the plant include leaves, stems, roots, buds, fruits, and seeds and may differ among developmental stages or vary seasonally. Three categories of plant tissue – meristem, vascular, and ground – have been reported as sites of ingestion or damage. Dermal tissue is not known to function as a nutrient source, but stomata are used by tingids and thaumastocorids as entry sites to reach leaf mesophyll (Buntin et al. 1996; Santadino et al. 2017). Cambium (lateral meristem) of pine and larch is fed upon by at least one species of aradid (Schuh and Slater 1995). However, ground and vascular tissues are far more common feeding sites for Heteroptera.

Salivary sheath feeders ingesting from vascular tissues are most likely to be found on stems and leaves, although roots or reproductive structures may also be used as feeding sites. Preference for stems and young shoots is seen in osmotic pump feeders, which extract nutrition from both phloem sieve cells and phloem parenchyma (Miles and Taylor 1994). The ground tissue most frequently used as a food source by cell-rupture feeders is parenchyma, including mesophyll, stem and root parenchyma, fruit pericarp, and seed endosperm and cotyledons. A given plant part may be used (and damaged) quite differently by vascular or cell-rupture feeders. Plant target tissue is therefore a more meaningful basis for comparison of feeding among Heteroptera than plant structure, and EPG is ideally suited for determining such behaviors.

3.2.1 Xylem Ingestion

Reliance on xylem as the sole source of nutrient acquisition is restricted to a small group of auchenorrhynchan insects comprising cicadas, cicadelline leafhoppers (sharpshooters), and froghoppers (also known as spittlebugs). However, use of xylem as a supplemental ingestion site, either for additional nutrients (Ebert et al. 2018) or hydration (Spiller et al. 1990), has been documented for Sternorrhyncha (e.g., aphids, Spiller et al. 1990; psyllids, Pearson et al. 2014; and whiteflies, Milenovic et al. 2019), phloem-feeding Auchenorrhyncha (Wayadande 1994, Chuche et al. 2017), coreids (Neal 1993; Mitchell 2006; Rodrigues et al. 2007; Maskey 2010), blissids (Backus et al. 2013), and in all species of pentatomids studied to date (Lucini and Panizzi 2018a; Mitchell et al. 2018a; Serteyn et al. 2020a). In aphids, xylem ingestion increases with starvation/dehydration time prior to EPG recording, indicating that it serves to maintain water balance (Spiller et al. 1990). In both Sternorrhyncha and Auchenorrhyncha, xylem feeding is considered to be a natural behavior (Pompon et al. 2011; Chuche et al. 2017), rather than simply a response to the stress of EPG recording in the laboratory. Histological evidence of penetration to xylem by coreids on wild host plants in the field (Mitchell 2006) supports this conclusion for Heteroptera as well.

Cell-rupture feeders, especially those feeding on seed endosperm, require a source of hydration to produce the large quantity of watery saliva needed to liquefy the food enzymatically (Miles 1972). Duration of xylem ingestion by *Nezara viridula* (L.) (Pentatomidae) was significantly longer when the insects fed on developing soybean pods than when they were offered only vegetative tissue and performed no cell-rupture feeding (Mitchell et al. 2018a). Furthermore, in most pentatomids, xylem ingestion precedes ingestion of seed endosperm (Lucini and Panizzi 2018a), suggesting that cell-rupture feeding requires advance hydration. First instar *N. viridula* ingest from xylem on leaves; in the absence of plant material, nymphs will imbibe from water-soaked cotton (Rivera and Mitchell 2020). Although no cell-rupture feeding occurs during this stadium (Rivera and Mitchell 2020), nymphs do obtain symbionts smeared by the female on the surface of the egg mass (Prado et al. 2006) and may expend watery saliva for this purpose.

Fluid in the xylem is under negative pressure (tension); thus, ingestion requires active muscular pumping prior to swallowing (Dugravot et al. 2008). Heteroptera lack the impressive cibarial pump development associated with dedicated xylem feeders, but nonetheless are able to maintain xylem ingestion for long periods of time: >1 h in fifth instar *Halyomorpha halys* Stål (Pentatomidae) and adult *Edessa meditabunda* (F.), *Euschistus heros* (F.), and *Diceraeus (Dichelops) furcatus* (F.) (Pentatomidae); >2 h in first instar *N. viridula* (Lucini and Panizzi 2018a, b; Serteyn et al. 2020b; Rivera and Mitchell 2020). The percentage of time devoted to xylem feeding is relatively consistent among adult pentatomids feeding on seed endosperm, ranging from 13% in *D. furcatus* to 20.7% in *N. viridula*, but is always less than that spent in cell-rupture feeding in the endosperm itself (Lucini and Panizzi 2018a; Mitchell et al. 2018a). In contrast, xylem-feeding activities exceeded time spent in phloem feeding for *Blissus insularis* Barber (Blissidae) on one susceptible variety of St. Augustine grass (Rangasamy et al. 2015). Dominance of the xylem ingestion behavior was also observed for *E. meditabunda* on soybean stems (Lucini and Panizzi 2016).

3.2.2 Phloem Ingestion

Many species of Heteroptera have been misidentified as phloem feeders due to careless interpretation of the term "plant sap" which is used indiscriminately in the older literature to refer to both plant cell contents in general and contents of the sieve elements specifically. Although phloem tissue provides the primary nutrient source for true bugs in several families, this mode of feeding is not dominant as in Sternorrhyncha or Auchenorrhyncha. Two very different heteropteran feeding strategies permit extraction of nutrients from phloem tissue. Salivary sheath feeding involves direct insertion of the stylet tips into sieve elements and ingestion of the translocated materials (Backus et al. 2005), whereas the osmotic pump strategy induces both phloem parenchyma and sieve elements to empty their contents into the apoplast (i.e., intercellular space), whence it is ingested (Miles and Taylor 1994).

Osmotic pump feeders may be recognized by the presence of a sucrase (α-glucosidase) in the watery saliva; salivary sheaths end in or near the phloem parenchyma rather than in the sieve tubes (Miles and Taylor 1994). No EPG study to date has examined this feeding mode, which has been associated with species in four tribes of Coreidae (Amorbini, Coreini, Dasynini, and Mictini) primarily in Australia but may be more widespread. Typical damage associated with the osmotic pump strategy includes the formation of sunken lesions beyond the internal reach of the stylets and wilting of the shoot towards the tip past the point of stylet insertion (Miles 1987; Miles and Taylor 1994; Taylor 1995; Hori 2000).

Observations that have been used to infer direct ingestion from sieve elements by Heteroptera include EPG, transmission of phloem-limited pathogens, and trophobiosis (ant tending); histological studies showing precise placement of stylets in the sieve cells have rarely been reported. Phloem ingestion through salivary sheath feeding has been identified in three superfamilies: Lygaeoidea (Blissidae, Piesmatidae, Lygaeidae), Pentatomoidea (Pentatomidae, Plataspidae), and Coreoidea (Coreidae). In Blissidae and Plataspidae, this mode of feeding is likely to be a family characteristic, but it is relatively rare among Pentatomidae, Lygaeidae, and Coreidae. Members of only three pentatomid tribes – Discocephalini, Ochlerini (Discocephalinae), and Edessini (Edessinae) – are thought to use phloem as a preferred source of nutrition, based on honeydew production and ant tending (Silva and Fernandes 2016).

Piesmatidae are considered by Cobben (1978) to be among the "most specialized sap feeders in the Heteroptera." This group has not been examined using EPG, but one species is known to transmit a virus of beets and a phloem-limited bacterium (Mitchell 2004) (see Sect. 3.4), suggesting an association with the phloem. However, leaf parenchyma cells are also damaged by piesmid feeding (Proeseler 1980). Chinch bugs (Blissidae: *Blissus* spp.), feeding on grasses, block sieve tubes and xylem with saliva; EPG recordings of two species have shown that ingestion occurs from both phloem and xylem (Backus et al. 2013). Early work by Miles (1959) showed flexibility in feeding by the lygaeid *Oncopeltus fasciatus* (Dallas), which uses cell rupture feeding on developing and mature seeds of milkweed but ingests from phloem using the salivary sheath strategy.

Seven species belonging to three coreid tribes (Amorbini, Cloresmini, and Colpurini) have been observed in trophobiotic relationships (Maschwitz et al. 1987; Silva and Fernandes 2016), possibly indicating direct feeding on phloem sap. An osmotic pump strategy would not likely produce the pure sugar-laden honeydew attractive to ants. However, no analysis of honeydew has been performed for species known to use the latter mode of feeding. Further research is needed to fully understand coreid feeding habits.

In Plataspidae, seven species are known to be tended by ants, which cluster around feeding bugs, protecting them from predators and ingesting droplets of anal exudate (Maschwitz et al. 1987; Waldkircher et al. 2004; Silva and Fernandes 2016). The long, coiled stylets of these plataspids penetrate the woody bark of trees to reach the phloem (Waldkircher et al. 2004). Other phloem-feeding plataspid species, such as the kudzu bug *Megacopta cribraria* (F.), feed on vegetative tissues of herbaceous plants. Histological evidence coupled with EPG identified phloem ingestion in *M. cribraria* (Stubbins et al. 2017).

Phloem ingestion has been established on the basis of EPG and histological evidence of stylet penetration in *E. meditabunda* (Pentatomidae: Edessinae) (Lucini and Panizzi 2016) and observations of sugary honeydew production and facultative ant-tending in *Edessa rufomarginata* De Geer (Silva and Fernandes 2016). Similar observations of ant-tending in two species of Discocephalinae (Guerra et al. 2011, Campos 2018), coupled with evidence of trypanosome transmission by a number of *Lincus* spp. (Discocephalinae: Ochlerini) (Mitchell et al. 2018b), suggest that phloem feeding could be characteristic of these two subfamilies. Phloem ingestion occurs sporadically in other pentatomid taxa in combination with cell-rupture type damage. *Palomena angulosa* Motschulsky (Pentatominae: Nezarini) ingests from phloem of leaves and stems, but nymphs also damage leaf parenchyma and flower buds (Hori 1984). Putative phloem feeding in addition to cell rupture feeding in mesophyll has been noted for *H. halys* (Pentatominae: Cappeini) based on EPG recordings (Serteyn et al. 2020a); this species also extensively damages fruits (Lee et al. 2013).

The ability of pentatomomorphan bugs to alternate between salivary sheath feeding in vascular tissues and cell rupture feeding in parenchyma may be more widespread than previously envisioned. Stylets of some mirid species terminate near vascular bundles, but ingestion of phloem is attributed to "chance or incidental feeding" (Wheeler 2001); no mirid has been shown to ingest directly from sieve elements as the primary mode of feeding (Wheeler 2001).

3.2.3 Endosperm Ingestion

Seeds provide a source of nutrition for many species of true bugs in the Pentatomomorpha; cell rupture feeding in endosperm and cotyledonary tissues releases the stored nutrients. Although preferences for developing or mature seeds have been noted for various taxa, few species use one or the other stage exclusively. Depending on the host plant, fruit pericarp surrounding the developing seed also may provide an additional source of nutrients during seed development (see Sect. 3.2.4).

Seed feeding is found in all superfamilies of Pentatomomorpha with the exception of the (primarily mycophagous) Aradoidea (Schuh and Slater 1995). The majority of phytophagous Lygaeoidea feed on mature seeds; hence the name "seed bugs" for Lygaeidae sensu lato (Sweet 1960, 2000). Pyrrhocoridae (Pyrrhocoroidea) feed on developing and mature seeds and the related Largidae on "seeds and plant juices" (Schuh and Slater 1995). Alydidae and Rhopalidae (Coreoidea) feed on reproductive plant parts including developing and mature seeds and fruits (Panizzi et al. 2000; Schaefer and Kotulski 2000). In the Coreidae, this habit is less common, but characteristic of at least two tribes, Anisoscelini and Clavigrallini (Mitchell 2000). Most stink bugs (Pentatomoidea: Pentatomidae) prefer developing seeds and fruits (Panizzi and Lucini 2017) and may be found with coreoids on the same crops, particularly legumes and tree fruits and nuts, forming economically damaging pest complexes (Mitchell 2000).

Two of the four described tactics of cell-rupture feeding (Backus et al. 2005, 2007) are used by true bugs consuming seed contents: lacerate-and-flush and macerate-and-flush. These tactics are now considered to represent the ends of a possible spectrum of feeding behaviors involving stylet movement and enzyme production; an additional tactic, lacerate/macerate-and-flush, has been proposed to describe such intermediate feeding (Lucini and Panizzi 2018a). All three feeding tactics result in degradation of the seed contents (primarily endosperm or cotyledons) from some combination of physical damage from stylet movement and chemical action of enzymes in the watery saliva, such that the semi-liquefied material may be withdrawn into the stylets. Reduced germination and seed quality, seed abortion, or even loss of entire fruiting structures may result from cell-rupture feeding (see Sect. 3.3).

3.2.4 Parenchyma Ingestion

Chlorotic spots on leaves, typical of tingid damage, result from removal of individual palisade mesophyll cell contents (Buntin et al. 1996); *Thaumastocoris peregrinus* Carpintero & Dellapé (Thaumastocoridae: Thaumastocorinae) apparently feed in a similar puncture-and-suck manner, removing chloroplasts from chlorenchyma cells (Santadino et al. 2017). The presence of uncollapsed cell walls in histological preparations after tingid or thaumastocorid feeding (e.g., Buntin et al.

1996; Santadino et al. 2017; Ishihara and Kawai 1981) and the termination of stylets in individual cells (Pollard 1959) suggest that this feeding mode may be distinct from other, more destructive cell-rupture tactics; for example, the empty cavities and degraded cell walls that are associated with mirid damage (Wheeler 2001). Nonetheless, heavy infestations of tingids on leaves can cause loss of foliage color and decreased photosynthetic activity as adjoining chlorotic spots coalesce (Johnson and Lyon 1994; Buntin et al. 1996). Royal palm bugs (Thaumastocoridae: Xylastodorinae) extract the contents of cells on recently opened palm fronds, creating small spots (Hill and Schafer 2000); this may also represent puncture-and-suck feeding on individual cells as in Thaumastocorinae.

Cell-rupture feeders using lacerate- or macerate-and-flush tactics may be found attacking parenchyma tissues of leaves, stems, roots, fruits, and seedlings. Little research has been done with root feeders, but the cydnid *Cyrtomenus bergi* Froeschner feeds on starch in the root parenchyma of cassava, destroying cell walls and cellular components in the process (Riis et al. 2003). Stem and leaf parenchyma and pods are targeted by some mirids, creating lesions that extend beyond the reach of the stylets and empty parenchyma cells while leaving cell walls temporarily intact but weakened (Miles 1987). A salivary pectinase is presumed to be responsible for this macerate-and-flush feeding (Miles and Taylor 1994). Other mirids rely more on a combination of stylet movements and salivary enzymes, disrupting leaf tissue with a series of slow stylet movements, followed by salivation and ingestion (Cervantes et al. 2017).

The majority of pentatomids attack reproductive plant parts (Panizzi and Lucini 2017), but some species, like the harlequin bug, *Murgantia histrionica* (Hahn), and the bagrada bug, *Bagrada hilaris* (Burmeister) are primarily leaf- and shoot-feeders (Blatchley 1926; Palumbo and Natwick 2010; Torres-Acosta et al. 2017). Cell-rupture feeding is apparently the source of damage. Other species of pentatomids will switch to stem parenchyma as necessary, although developing seeds and fruits remain preferred. Adult cabbage bugs, *Eurydema rugosa* Motschulsky (Pentatomidae), feed on mesophyll and stem parenchyma in the absence of reproductive tissue but switch exclusively to buds and pods when available (Hori 1968). EPG recordings of the pentatomid *D. furcatus* on wheat indicate that cell rupture feeding occurs on both stems and ears (Lucini and Panizzi 2017a). In contrast, neither *N. viridula* nor *Piezodorus guildinii* (Westwood) use stem parenchyma of their host plants as a nutrient source; cell-rupture feeding is restricted to reproductive tissues. When these bugs feed on stems, the target ingestion tissue is xylem (Lucini et al. 2016; Mitchell et al. 2018a).

Feeding by true bugs on seedlings has been reported in the economic literature, associated especially with no-tillage systems (Sedlacek and Townsend 1988; Lucini and Panizzi 2017b). Several species that primarily attack seeds also have been reported to damage seedlings, often those of entirely different crop plants. For example, the lygaeid *Nysius huttoni* White damages wheat grains but also produces cankerous growths and girdling in seedlings of *Brassica* spp. and may kill the plants (Tiwari and Wratten 2019). Similarly, *Euschistus* spp. (Pentatomidae), serious pests of reproductive-stage soybean, tree fruits, pecans, and cotton (McPherson and

McPherson 2000), cause lesions, tillering, wilting, stunting, and death of early-stage corn when feeding at the base of the young plants (Sedlacek and Townsend 1988). Although wilting and girdling may be the result of blocked vascular tissue from salivary sheath feeding, cell-rupture feeding is more likely to produce lesions and could be responsible for all of the symptoms observed. EPG recordings of *Diceraeus (Dichelops) melacanthus* Dallas on maize seedling stems indicate cell-rupture feeding as the cause of damage (Lucini and Panizzi 2017b).

Pericarp parenchyma of fruits and developing pods serves as a food source for a wide variety of true bugs, including mirids, coreoids, and pentatomids. Endosperm of developing seeds may be targeted as well, if fruit diameter and stylet length permit, but in many cases only the soft tissue of the ovary wall is fed upon, disfiguring the fruit and altering the consistency and taste (e.g., Kurian et al. 1976; Handley and Pollard 1993; Panizzi and Alves 1993; Dreyer et al. 1994; Coombs and Khan 1998; Wheeler 2001).

3.2.5 Inflorescence and Bud Feeding

Mirids are most closely associated with feeding on inflorescences. Bud and flower structures are fed upon and developing anthers and their pollen sacs are used preferentially by some species (Wheeler 2001). Even mature pollen may provide a source of nutrition, although entire grains are too large to be taken up in the food canal. Wheeler (2001) examined possible mechanisms for mirids to obtain access to the contents of pollen grains and suggested that a combination of enzymatic action and nutrient diffusion may be responsible.

Bud feeding has been extensively studied in conjunction with damage to cotton squares (flower buds) by *Lygus* bugs and other pest mirids. Pollen sacs are the preferred feeding site on cotton (Mauney and Henneberry 1979). *Lygus hesperus* Knight perform multiple-cell laceration on squares accompanied by salivation; after cell contents are macerated by enzymatic activity, ingestion follows (Cline and Backus 2002; Backus et al. 2007). Squares on which *L. hesperus* have previously fed are more attractive than control squares; this was attributed to continued action of salivary enzymes (Cooper and Spurgeon 2013). EPG recordings of adult and third-instar *Lygus* spp. feeding on pinhead squares of cotton confirmed cell-rupture feeding (Cervantes et al. 2016, 2019); additional EPG coupled with histological studies showed cell rupture and ingestion by *L. lineolaris* (Palisot de Beauvoir) to occur in the staminal column (Cervantes et al. 2017). However, *Creontiades dilutus* (Stål) targeted ovules of alfalfa rather than male reproductive structures, entering the flower through the sepals and leading to flower drop; feeding also occurred on stems and peduncles (Hori and Miles 1993).

3.3 Damage to Plants by True Bugs

The extent of plant damage caused by true bugs depends not only on the plant part attacked, but the feeding strategy and tactics employed and the plant response. The effect of the latter two aspects has been elegantly described for hopperburn, caused by the auchenorrhynchan *Empoasca fabae* (Harris) (Backus et al. 2005). According to this model, the plant reacts to hopperburn initiation (i.e., insect feeding behaviors, including salivation and stylet movements) with physiological, biochemical, and molecular wound responses, collectively termed the hopperburn cascade (Backus et al. 2005).

Understanding the plant response to any piercing-sucking feeder thus requires identification of the feeding strategy employed and the components of the watery saliva. A detailed review of plant wound responses is beyond the scope of this chapter; following the model of Backus et al. (2005) for hopperburn, our objective is to focus primarily on initiation behaviors by the insect.

3.3.1 Damage from Salivary Sheath Feeding

Due to the precise and delicate placement of stylets in the vascular tissue, this feeding strategy does not always produce extensive visible symptoms; nonetheless, damage may be severe. Wilting of squash seedlings caused by *Anasa tristis* (De Geer) (Coreidae) has been attributed to interruption of xylem transport, either by blockage of vessels with gelling saliva or physical damage to the cells eliciting a plant wound response (Neal 1993). Complete girdling of the stem, with concomitant damage to all xylem vessels, is necessary to induce wilting and unlikely to occur except under high population pressure (Neal 1993). Salivary sheath feeding in the phloem drains the plant of needed photosynthate, resulting in reduced growth, yield loss in crops, and even plant death.

Kudzu bugs (Plataspidae) on soybean at high densities reduce seed weight and seeds per pod (Seiter et al. 2013). Chinch bugs feeding on corn and grain sorghum reduce the supply of photosynthates to the plant and may plug xylem and phloem with salivary deposits (Painter 1928). The stem-feeding pentatomid *E. meditabunda* is reported to kill potato plants (Rizzo 1971) and can reduce soybean yield, although not as severely as seed-feeding pentatomids (Silva et al. 2012; Husch et al. 2014). Phloem-feeding by pentatomids poses an additional problem due to transmission of pathogenic trypanosomatids (see Sect. 3.4).

3.3.2 Damage from Osmotic Pump Feeding

To date, only a handful of coreid species have been definitively identified as osmotic pump feeders, and of that group, few are crop pests of sufficient importance to be studied in depth. Miles (1987) originally described the damage from this mode of feeding by *Amblypelta* sp. as a darkened lesion, not always visible externally. On young sweet potato and cassava stems, these feeding lesions, when superficially visible, were noted as "water-soaked," and were characterized internally by collapsed cells and enlarged intercellular spaces, creating sunken areas on the stem surface (Miles 1987). Subsequent research (Miles and Taylor 1994) showed that the stem lesions resulted in wilting beyond the point of damage; indeed, several of the coreids identified as probable osmotic pump feeders are known colloquially as "tip-wilters." Within 48 h of feeding by *Mictis profana* (F.) on stems of *Acacia iteaphylla* F. Muell. ex Benth, the leaves beyond the lesion become black, twisted, and dead, and the stem collapses (Taylor 1995).

Amorbus spp. are pests of *Eucalyptus* plantations; *A. obscuricornis* (Westwood) similarly produces wilting and necrosis of young apical shoots and reduces tree growth (Steinbauer et al. 1997). Curiously, *Gelonus tasmanicus* (Le Guillou), which possesses the salivary enzymes characteristic of osmotic pump feeders, feeds on stems and mature leaves but causes no shoot wilting; the quantity of saliva, duration of feeding, and the maturity of the tissue attacked could explain this difference in damage (Steinbauer et al. 1997). In his extensive study of the biology of Australian coreids, Kumar (1966) notes that *Amorbus* spp. and *Aulacosternum nigrorubrum* Dallas [another osmotic pump feeder; Taylor (1995)] are strictly sap (i.e., phloem) feeders, whereas *Mictis* and *Amblypelta* are capable of tip-wilting but also damage fruit. It is unknown whether this flexibility represents a change in feeding strategy or the application of osmotic pump feeding to fruit parenchyma tissue.

3.3.3 Damage from Cell-Rupture Feeding

Lacerate- and macerate-and-flush feeding tactics are both highly damaging, destroying pockets of cells and creating lesions in parenchyma tissues and emptying seeds of their contents. Pod and fruit abscission may result from damage to young reproductive structures. Germination rates of surviving seeds are reduced, and grain quality is diminished (e.g., Leigh et al. 1988; Depieri and Panizzi 2011; Tuelher et al. 2020). Mirids produce lesions on cocoa pods and mangos and spots on cotton bolls and induce pod abscission in legumes, formation of callus tissue in apples, scarring in peaches, and fruit malformation in pears, apples, and avocados (Wheeler 2001). Similar damage to peaches, apples, and pears is caused by various pentatomids (McPherson and McPherson 2000). Coreids in the genus *Leptoglossus* Guérin-Méneville attack peaches, pomegranate, guava, and various citrus fruits; juice vesicles of the latter are damaged (Mitchell 2000). Economic damage by mirids is extensively reviewed by Wheeler (2001); for detailed information on damage by cell-rupture feeding Pentatomomorpha, the reader may consult Schaefer and Panizzi (2000).

The most detailed analysis of cell-rupture feeding damage focuses on the effects of mirids, especially *Lygus* bugs, on cotton. Feeding by *L. hesperus* and *Pseudatomoscelis seriatus* (Reuter) causes necrosis of the staminal column in small (pinhead) squares and the anthers in larger flower buds; the ovary and pedicel may also be damaged (Mauney and Henneberry 1979). Histological studies (Williams and Tugwell 2000) showed damage mainly to male structures. Cells were disorganized and degraded, cell walls were fragmented, and after extended feeding, anthers were shrunken and the staminal column was destroyed (Williams and Tugwell 2000). Macerate-and-flush damage to the staminal column induced the production of tannins, presumably as a plant response to wounding (Cervantes et al. 2017). Even a single cell-rupture event could produce tannin deposition, but the amount of tannin increased with additional feeding. Presence of tannins did not deter feeding, and the bugs consumed the tannins along with the cell contents during subsequent ingestion (Cervantes et al. 2017).

Another form of cell-rupture feeding, sequential penetration of individual cells (i.e., puncture-and-suck), is performed by tingids and thaumastocorines. These bugs feed predominantly on mature leaves, causing symptoms ranging from chlorotic spots to leaf silvering, bronzing, and abscission. Ornamental trees and shrubs become disfigured, and in the case of eucalyptus attacked by *T. peregrinus*, tree death may result from high infestations (Jacobs and Nesser 2005, cited in Santadino et al. 2017). Stylet entry is through the stomata in *T. peregrinus* and many tingid species studied, but movement of the tingid stylet bundle beneath the epidermis is intracellular (Pollard 1959) and may cause damage en route to the target tissue. Published photomicrographs (Ishihara and Kawai 1981; Buntin et al. 1996; Santadino et al. 2017) indicate that individual, adjacent palisade parenchyma cells are emptied of their contents (including chloroplasts), resulting in chlorosis. Leaf feeding injuries from the tingid *Stephanitis pyrioides* (Scott) on azalea reduce the rate of photosynthesis due to loss of chlorophyll and interference with stomatal action (Buntin et al. 1996), but azalea plantings can withstand a relatively high level of infestation and canopy damage before overall growth is affected (Klingeman et al. 2001).

3.4 Acquisition and Inoculation of Pathogens During Feeding

Heteroptera have been historically unappreciated as vectors of plant pathogens, although heteropteran species have been implicated in the transmission of most groups of pathogenic organisms, with the exception of spiroplasmas and fastidious xylem-limited bacteria. Fungal transmission is non-circulative, but involvement of the salivary glands in circulative, propagative, persistent transmission has been documented for other types of pathogens.

Few instances of virus transmission are associated with Heteroptera. *Piesma quadratum* (Fieber) (Piesmatidae) transmits the rhabdovirus that causes beet leaf curl disease (Proesler 1980). Virus particles are acquired from the phloem parenchyma and are found in the midgut, hemolymph, and salivary glands; individuals remain inoculative permanently (Mitchell 2004 and references therein). A mirid, *Engytatus nicotianae* (Koningsberger) transmits *Velvet tobacco mottle virus* in a completely different manner, involving the midgut rather than the salivary glands, termed ingestion/defecation (Gibb and Randles 1991).

Transmission of phytoplasmas is propagative, persistent, and assumed to be dependent on phloem feeding by the vector. Auchenorrhyncha are the most common vectors (Weintraub and Wilson 2010). However, two species of true bugs have been reported to transmit phytoplasmas, and surveys using PCR detection methods have unearthed several additional potential heteropteran vectors (Mitchell 2004). The brown marmorated stink bug, *H. halys*, is the vector of Paulownia witches'-broom disease of empress trees in East Asia (Hiruki 1999). Recent EPG recordings of *H. halys* indicate putative phloem ingestion in addition to the more common cell-rupturing activity of this bug (Serteyn et al. 2020a); the latter mode of feeding would be highly unlikely to successfully inoculate a phloem-limited pathogen. Transmission of root wilt in coconut has been attributed to a lace bug, *Stephanitis typica* (Distant). Tingids generally feed by penetrating leaf palisade mesophyll cells, but stylets of this species have been observed to terminate in phloem (Mathen et al. 1990). Phytoplasma transmission has always been associated with piercing-sucking insects that feed in a non-destructive manner, without damaging conductive tissues (Weintraub and Beanland 2006); i.e., salivary sheath feeding, but transmission may not be limited to this feeding mode.

Successful, effective inoculation of phloem-limited pathogens into a sieve tube requires penetration without major cell damage; smaller Heteroptera could conceivably do this, but the stylets of larger bugs such as *A. tristis* would likely cause extensive cellular disruption (A. C. Wayadande, Oklahoma State University, pers. comm. to PLM). *Serratia marcescens* Bizio, the phloem-colonizing bacterium causing cucurbit yellow vine disease (Bruton et al. 2003), is propagative within *A. tristis*, but it is presently unknown whether the bug transmits the bacterium directly into sieve elements or to the phloem by an as-yet-unknown mechanism (Wayadande et al. 2005). Like the lace bug and the stink bug reported to transmit phytoplasmas, *A. tristis* is not a dedicated phloem feeder; squash bugs ingest from cucurbit fruit as well as leaves and stems (Doughty et al. 2016).

Stylet insertion holes provide entry for many non-fastidious bacterial plant pathogens, and the incidental spread of disease in this manner has been associated with a variety of heteropteran species. Mitchell (2004) lists 27 species in four families (Miridae, Coreidae, Pentatomidae, Pyrrhocoridae) associated with bacterial diseases of plants, but the bulk of these reports lack actual confirmation of transmission. One notable exception is the relationship between *Pantoea agglomerans*, which causes an internal boll rot of cotton, and the pentatomid *N. viridula*, which was shown to be a competent vector of the bacterium (Medrano et al. 2007). Even first instars could acquire the bacterium, from infected cut pieces of green bean (Esquivel and Medrano 2014). Acquisition by first instars likely occurred through xylem ingestion (Rivera and Mitchell 2020), but older nymphs and adults feed directly on boll tissue and inoculate *P. agglomerans* through cell rupture feeding.

Pentatomids, pyrrhocorids, and coreoids transmit fungal diseases to a variety of crops, including coffee, cotton, rice, legumes, tomato, citrus, and pistachio. In the majority of cases, the pathogen is the yeast *Eremothecium coryli* (Peglion) Kurtzman, which has a close relationship especially with Pentatomidae (Mitchell 2004; Mitchell et al. 2018b). Like *P. agglomerans*, this yeast is found in the rostrum, head, and alimentary canal of *N. viridula* (Esquivel and Medrano 2012); other bacteria present only in the alimentary canal are not transmitted by *N. viridula*. On legumes, *Chinavia hilaris* (Say) (Pentatomidae) adults can retain *E. coryli* for up to 60 days, but fifth instars lose infectivity following the molt (Clark and Wilde 1970), indicating a semi-persistent mode of transmission.

A wide variety of heteropteran species harbor plant-inhabiting trypanosomatids in the genus *Phytomonas*, but only a few of these cause disease (Camargo and Wallace 1994; Camargo 1999; Godoi et al. 2002). Trypanosomatids are acquired by cell-rupture feeders from various fruits, by latex- and phloem-feeders from the laticifers of euphorbs and other latex-producing plants, and by phloem ingestion from oil palms and coconut (Dollett 2016) and are found in the digestive tract and salivary glands of the insect host (Camargo and Wallace 1994). The phloem-restricted trypanosomatids are the only pathogenic forms, causing sudden wilt of oil palms, hartrot of coconut, and phloem necrosis of coffee (Camargo 1999). In each case where a vector has been conclusively identified for the disease, it has been a phloem-feeding pentatomid in the genus *Lincus* Stål (Dollett 2016), although an unidentified species of *Ochlerus* Spinola (Pentatomidae) is a suspected vector of hartrot (Dollett 2016). Cell-rupture feeding Heteroptera, although capable of transmitting fruit-inhabiting *Phytomonas*, are unlikely to be of any economic importance.

3.5 Concluding Remarks

The variety of feeding modes associated with heteropteran taxa, the ability of many true bugs to employ more than one feeding strategy, the poorly explored potential for pathogen transmission, and especially the use of – and damage to – virtually all plant structures and tissues makes the study of Heteroptera-plant interactions a definite challenge. Clearly, many unanswered questions remain. How widespread is the osmotic pump strategy? How do large heteropterans inoculate sieve elements without damaging the phloem? What determines switching behavior between cell-rupture and salivary sheath feeding, or between phloem- and xylem-ingestion, and what limits this flexibility? Future EPG studies, coupled with more traditional behavioral research methods, will help to resolve these and other questions and expand our understanding of true bug feeding and causes of damage.

References

Backus EA, Serrano MS, Ranger CM (2005) Mechanisms of hopperburn: an overview of insect taxonomy, behavior, and physiology. Annu Rev Entomol 50:125–151

Backus EA, Cline AR, Ellerseick MR, Serrano MS (2007) *Lygus hesperus* (Hemiptera: Miridae) feeding on cotton: new methods and parameters for analysis of nonsequential electrical penetration graph data. Ann Entomol Soc Am 100:296–310

Backus EA, Rangasamy M, Stamm M, McAuslane HJ, Cherry R (2013) Waveform library for chinch bugs (Hemiptera: Heteroptera: Blissidae): characterization of electrical penetration graph waveforms at multiple input impedances. Ann Entomol Soc Am 106:524–539

Blatchley WS (1926) Heteroptera or true bugs of eastern North America. Nature Publication, Indianapolis

Bruton BD, Mitchell F, Fletcher J, Pair SD, Wayadande A, Melcher U, Brady J, Bextine B, Popham TW (2003) *Serratia marcescens*, a phloem-colonizing, squash bug-transmitted bacterium: causal agent of cucurbit yellow vine disease. Plant Dis 87:937–944

Buntin GD, Braman SK, Gilbertz DA, Phillips DV (1996) Chlorosis, photosynthesis, and transpiration of azalea leaves after azalea lace bug (Heteroptera: Tingidae) feeding injury. J Econ Entomol 89:990–995

Camargo EP (1999) *Phytomonas* and other trypanosomatid parasites of plants and fruit. Adv Parasitol 42:29–112

Camargo EP, Wallace FG (1994) Vectors of plant parasites of the genus *Phytomonas* (Protozoa, Zoomastigophorea, Kinetoplastida). In: Harris KF (ed) Advances in disease vector research, vol 10. Springer, Heidelberg, pp 333–359

Campos LA (2018) Hidden diversity of Discocephalinae (Pentatomidae): current knowledge and perspectives. Conference: sixth quadrennial meeting of the International Heteropterists' Society. La Plata, Argentina, December 2018. https://doi.org/10.13140/RG.2.2.15270.16969

Cervantes FA, Backus EA, Godfrey L, Akbar W, Clark TL (2016) Characterization of an EPG waveform library for adult *Lygus lineolaris* and *Lygus hesperus* (Hemiptera: Miridae) feeding on cotton squares. Ann Entomol Soc Am 109:684–697

Cervantes FA, Backus EA, Godfrey L, Wallis C, Akbar W, Clark TL, Rojas MG (2017) Correlation of electropenetrography waveforms from *Lygus lineolaris* (Hemiptera: Miridae) feeding on cotton squares with chemical evidence of inducible tannins. J Econ Entomol 110:2068–2075

Cervantes FA, Backus EA, Godfrey L, Rojas MG, Akbar W, Clark TL (2019) Quantitative differences in feeding behavior of *Lygus lineolaris* (Hemiptera: Miridae) on transgenic and non-transgenic cotton. J Econ Entomol 112:1920–1925

Chuche J, Sauvion N, Tiéry D (2017) Mixed xylem and phloem sap ingestion in sheath-feeders as normal dietary behavior: evidence from the leafhopper *Scaphoideus titanus*. J Insect Physiol 102:62–72

Clark RG, Wilde GE (1970) Association of the green stink bug and the yeast spot disease organism of soybeans. I. Length of retention, effect of molting, isolation from feces and saliva. J Econ Entomol 63:200–204

Cline AR, Backus EA (2002) Correlations among AC electronic monitoring waveforms, body postures, and stylet penetration behaviors of *Lygus hesperus* (Hemiptera: Miridae). Environ Entomol 31:543–549

Cobben RH (1978) Evolutionary trends in Heteroptera. Part II. Mouthpart-structures and feeding strategies. Mededelingen Landbouwhogeschool Wageningen 78-5:1–407

Coombs M, Khan SA (1998) Population levels and natural enemies of *Plautia affinis* Dallas (Hemiptera: Pentatomidae) on raspberry, *Rubus idaeus* L., in south-eastern Queensland. Aust J Entomol 37:125–129

Cooper WR, Spurgeon DW (2013) Response by *Lygus hesperus* (Hemiptera: Miridae) adults to salivary preconditioning of cotton squares. J Entomol Sci 48:261–264

Depieri RA, Panizzi AR (2011) Duration of feeding and superficial and in-depth damage to soybean seed by selected species of stink bugs (Heteroptera: Pentatomidae). Neotrop Entomol 40:197–203

Dollet M (2016) Heteroptera and plant trypanosomatids (*Phytomonas* spp.). In: Brown J (ed) Vector-mediated transmission of plant pathogens. APS Press, St. Paul, pp 379–397

Doughty HB, Wilson JM, Schultz PB, Kuhar TP (2016) Squash bug (Hemiptera: Coreidae): biology and management in cucurbitaceous crops. J Integr Pest Manage 7:1;1–8

Dreyer H, Baumgärtner J, Tamò M (1994) Seed-damaging field pests of cowpea (*Vigna unguiculata* L. Walp.) in Benin: occurrence and pest status. Int J Pest Manage 40:252–260

Dugravot S, Backus EA, Reardon BJ, Miller TA (2008) Correlations of cibarial muscle activities of *Homalodisca* spp. sharpshooters (Hemiptera: Cicadellidae) with EPG ingestion waveform and excretion. J Insect Physiol 54:1467–1478

Ebert TA, Backus EA, Shugart HJ, Rogers ME (2018) Behavioral plasticity in probing by *Diaphorina citri* (Hemiptera, Liviidae): ingestion from phloem versus xylem is influenced by leaf age and surface. J Insect Behav 31:119–137

Esquivel JF, Medrano EG (2012) Localization of selected pathogens of cotton within the southern green stink bug. Entomol Exp Appl 142:114–120

Esquivel JF, Medrano EG (2014) Ingestion of a marked bacterial pathogen of cotton conclusively demonstrates feeding by first instar southern green stink bug (Hemiptera: Pentatomidae). Environ Entomol 43:110–115

Gibb KS, Randles JW (1991) Transmission of velvet tobacco mottle virus and related viruses by the mirid *Cyrtopeltis nicotianae*. In: Harris KF (ed) Advances in disease vector research, vol 7. Springer, Heidelberg, pp 1–17

Godoi MMI, Serrano MG, Teixeira MMG, Camargo EP (2002) A PCR-based survey on *Phytomonas* (Euglenozoa: Trypanosomatidae) in phytophagous hemipterans of the Amazon region. J Eukaryot Microbiol 49:275–279

Guerra TJ, Camarota F, Castro FS, Schwertner CF, Grazia J (2011) Trophobiosis between ants and *Eurystethus microlobatus* Ruckes 1966 (Hemiptera: Heteroptera: Pentatomidae) a cryptic, gregarious and subsocial stinkbug. J Nat Hist 45:1101–1117

Handley DT, Pollard JE (1993) Microscopic examination of tarnished plant bug (Heteroptera: Miridae) feeding damage to strawberry. J Econ Entomol 86:505–510

Hill L, Schaefer CW (2000) Palm bugs (Thaumastocoridae). In: Schaefer CW, Panizzi AR (eds) Heteroptera of economic importance. CRC Press, Boca Raton, pp 139–142

Hiruki C (1999) Paulownia witches'-broom disease important in East Asia. Acta Hortic 496:63–68

Hori K (1968) Feeding behavior of the cabbage bug, *Eurydema rugosa* Motschulsky (Hemiptera: Pentatomidae) on the cruciferous plants. Appl Entomol Zool 3:26–36

Hori K (1984) Feeding site of *Palomena angulosa* Motschulsky (Hemiptera: Pentatomidae) on potato plants and injury caused by the feeding. Appl Entomol Zool 19:476–482

Hori K (2000) Possible causes of disease symptoms resulting from the feeding of phytophagous Heteroptera. In: Schaefer CW, Panizzi AR (eds) Heteroptera of economic importance. CRC Press, Boca Raton, pp 11–36

Hori K, Miles PW (1993) The etiology of damage to lucerne by the green mirid, *Creontiades dilutus* (Stål). Australian J Exp Agric 33:327–331

Husch PE, de Oliveira MCN, Sósa-Gomez DR (2014) Characterization of injury caused by *Edessa meditabunda* (F.), *Chinavia impicticornis* (Stål), and *Piezodorus guildinii* (West.) (Hemiptera: Pentatomidae) to soybean. Neotrop Entomol 43:276–281

Ishihara R, Kawai S (1981) Feeding habits of the azalea lace bug *Stephanitis pyrioides* Scott (Hemiptera: Tingidae). Jpn J Appl Entomol Zool 25:200–202 (*in Japanese*)

Jacobs DH, Nesser S (2005) *Thaumastocoris australicus* Kirkaldy (Heteroptera: Thaumastocoridae): a new insect arrival in South Africa, damaging to *Eucalyptus* trees. S Afr J Sci 101:233–236

Johnson WT, Lyon HH (1994) Insects that feed on trees and shrubs, 2nd edn, revised. Cornell University Press, Ithaca

Klingeman WE, Braman SK, Buntin SD (2001) Azalea growth in response to azalea lace bug (Heteroptera: Tingidae) feeding. J Econ Entomol 94:129–137

Kumar R (1966) Studies on the biology, immature stages, and relative growth of some Australian bugs of the superfamily Coreoidea (Hemiptera: Heteroptera). Aust J Zool 14:895–991

Kurian C, Abraham VA, Koya A (1976) A new enemy of coconut in India [*Paradasynus*, Coreidae]. Indian Farming 26:11–12

Lee D-H, Short BD, Joseph SV, Bergh JC, Leskey TC (2013) Review of the biology, ecology, and management *of Halyomorpha halys* (Hemiptera: Pentatomidae) in China, Japan, and the Republic of Korea. Environ Entomol 42:627–641

Leigh TF, Kerby TA, Wynholds PF (1988) Cotton square damage by the plant bug, *Lygus hesperus* (Hemiptera: Heteroptera: Miridae), and abscission rates. J Econ Entomol 81:1328–1337

Lucini T, Panizzi AR (2016) Waveform characterization of the soybean stem feeder *Edessa meditabunda* (F.) (Hemiptera: Heteroptera: Pentatomidae): overcoming the challenge of wiring pentatomids for EPG. Entomol Exp Appl 158:118–132

Lucini T, Panizzi AR (2017a) Probing behavior of *Dichelops furcatus* (F.) (Heteroptera: Pentatomidae) on wheat plants characterized by electropenetrography (EPG) and histological studies. J Insect Sci 17:65;1–15

Lucini T, Panizzi AR (2017b) Feeding behavior of the stink bug *Dichelops melacanthus* (Heteroptera: Pentatomidae) on maize seedlings: an EPG analysis at multiple input impedances and histology correlation. Ann Entomol Soc Am 110:160–171

Lucini T, Panizzi AR (2018a) Electropenetrography (EPG): a breakthrough tool unveiling stink bug (Pentatomidae) feeding on plants. Neotrop Entomol 47:1–13

Lucini T, Panizzi AR (2018b) Electropenetrography monitoring of the neotropical brown-stink bug (Hemiptera: Pentatomidae) on soybean pods: an electrical penetration graph-histology analysis. J Insect Sci 18:5;1–14

Lucini T, Panizzi AR, Backus EA (2016) Characterization of an EPG waveform library for redbanded stink bug, *Piezodorus guildinii* (Hemiptera: Pentatomidae), on soybean plants. Ann Entomol Soc Am 109:198–210

Maschwitz U, Fiala B, Dolling WR (1987) New trophobiotic symbioses of ants with South East Asian bugs. J Nat Hist 21:1097–1107

Maskey K (2010) Comparison of electrical penetration graph waveforms of squash bug feeding on watermelon and its relatives. MS thesis, Oklahoma State Univ, Stillwater, Oklahoma

Mathen K, Rajan P, Nair CPR, Sasikala M, Gunasekharan M, Govindankutty MP, Solomon JJ (1990) Transmission of root (wilt) disease to coconut seedlings through *Stephanitis typica* (Distant) (Heteroptera: Tingidae). J Tropic Agric 67:69–73

Mauney JR, Henneberry TJ (1979) Identification of damage symptoms and patterns of feeding of plant bugs in cotton. J Econ Entomol 72:496–501

McPherson JE, McPherson RM (2000) Stink bugs of economic importance in America north of Mexico. CRC Press, Boca Raton

Medrano EG, Esquivel JF, Bell AA (2007) Transmission of cotton seed and boll rotting bacteria by the southern green stink bug (*Nezara viridula* L.). J Appl Microbiol 103:436–444

Milenovic M, Wosula EN, Rapisarda C, Legg JP (2019) Impact of host plant species and whitefly species on feeding behavior of *Bemisia tabaci*. Front Plant Sci 10:1–14

Miles PW (1959) The salivary secretions of a plant-sucking bug, *Oncopeltus fasciatus* (Dall.) (Heteroptera: Lygaeidae). I – the types of secretions and their roles during feeding. J Insect Physiol 3:243–255

Miles PW (1972) The saliva of Hemiptera. Adv Insect Physiol 9:183–255

Miles PW (1987) Plant-sucking bugs can remove the contents of cells without mechanical damage. Experientia 43:937–939

Miles PW, Taylor GS (1994) 'Osmotic pump' feeding by coreids. Entomol Exp Appl 73:163–173

Mitchell PL (2000) Leaf-footed bugs (Coreidae). In: Schaefer CW, Panizzi AR (eds) Heteroptera of economic importance. CRC Press, Boca Raton, pp 337–403

Mitchell PL (2004) Heteroptera as vectors of plant pathogens. Neotrop Entomol 33:519–545

Mitchell PL (2006) Polyphagy in true bugs: a case study of *Leptoglossus phyllopus* (L.) (Hemiptera, Heteroptera, Coreidae). Denisia 19. Neue Serie 50:1117–1134

Mitchell PL, Cooke SB, Smaniotto LF (2018a) Probing behavior of *Nezara viridula* on soybean: characterization and comparison of electrical penetration graph (EPG) waveforms on vegetative and reproductive plant structures. J Agric Urban Entomol 34:19–43

Mitchell PL, Zeilinger AR, Medrano EG, Esquivel JF (2018b) Pentatomoids as vectors of plant pathogens. In: McPherson JE (ed) Invasive stink bugs and related species (Pentatomoidea): biology, higher systematics, semiochemistry, and management. CRC Press, Boca Raton, pp 611–640

Neal JJ (1993) Xylem transport interruption by *Anasa tristis* feeding causes *Cucurbita pepo* to wilt. Entomol Exp Appl 69:195–200

Painter RH (1928) Notes on the injury to plant cells by chinch bug feeding. Ann Entomol Soc Am 21:232–242

Palumbo JC, Natwick ET (2010) The bagrada bug (Hemiptera: Pentatomidae): a new invasive pest of cole crops in Arizona and California. Online Plant Health Progress. https://doi.org/10.1094/PHP-2010-0621-01-BR

Panizzi AR, Alves RML (1993) Performance of nymphs and adults of the southern green stink bug (Heteroptera: Pentatomidae) exposed to soybean pods at different phenological stages of development. J Econ Entomol 86:1088–1093

Panizzi AR, Lucini T (2017) Host plant-stink bug (Pentatomidae) relationships. In: Čokl A, Borges M (eds) Stink bugs: biorational control based on communication processes. CRC Press, Boca Raton, pp 31–58

Panizzi AR, Schaefer CW, Natuhara Y (2000) Broad-headed bugs (Alydidae). In: Schaefer CW, Panizzi AR (eds) Heteroptera of economic importance. CRC Press, Boca Raton, pp 321–336

Pearson CC, Backus EA, Shugart HJ, Munyaneza JE (2014) Characterization and correlation of EPG waveforms of *Bactericera cockerelli* (Hemiptera: Triozidae): variability in waveform appearance in relation to applied signal. Ann Entomol Soc Am 107:650–666

Pollard DG (1959) Feeding habits of the lace-bug *Urentius aegyptiacus* Bergevin (Hemiptera: Tingidae). Ann Appl Biol 47:778–782

Pompon J, Quiring D, Goyer C, Giordanengo P, Pelletier Y (2011) A phloem-sap feeder mixes phloem and xylem sap to regulate osmotic potential. J Insect Physiol 57:1317–1322

Prado SS, Rubinoff D, Almeida RPP (2006) Vertical transmission of a pentatomid caeca-associated symbiont. Ann Entomol Soc Am 99:577–585

Proeseler G (1980) Piesmids. In: Harris KF, Maramorosch K (eds) Vectors of plant pathogens. Academic, New York, pp 97–113

Rangasamy M, McAuslane HJ, Backus EA, Cherry RH (2015) Differential probing behavior of *Blissus insularis* (Hemiptera: Blissidae) on resistant and susceptible St. Augustinegrasses. J Econ Entomol 108:780–788

Riis L, Bellotti AC, Bonierbale M, O'Brien GM (2003) Cyanogenic potential in cassava and its influence on a generalist insect herbivore *Cyrtomenus bergi* (Hemiptera: Cydnidae). J Econ Entomol 96:1905–1914

Rivera KL, Mitchell PL (2020) Feeding behavior of *Nezara viridula* first instars: EPG analysis and effect of food availability on subsequent development. J Agric Urban Entomol 36:47–63

Rizzo HFE (1971) Aspectos morphológicos y biológicos de *Edessa meditabunda* (F.) (Hemiptera, Pentatomidae). Rev Peruana Entomol 14:272–281

Rodrigues D, Sampaio DS, Isaias RMDS, Moreira GRP (2007) Xylem and seed feeding by two passion vine leaffooted bugs, *Holymenia clavigera* and *Anisoscelis foliacea marginella* (Hemiptera: Coreidae: Anisoscelini) with notes on mouthpart morphology. Ann Entomol Soc Am 100:907–913

Santadino MV, Brentassi ME, Fanello DD, Coviella C (2017) First evidence that *Thaumastocoris peregrinus* (Heteroptera: Thaumastocoridae) feeds from mesophyll of eucalyptus leaves. Environ Entomol 46:251–257

Schaefer CW, Kotulski J (2000) Scentless plant bugs (Rhopalidae). In: Schaefer CW, Panizzi AR (eds) Heteroptera of economic importance. CRC Press, Boca Raton, pp 309–320

Schaefer CW, Panizzi AR (eds) (2000) Heteroptera of economic importance. CRC Press, Boca Raton

Schuh RT, Slater JA (1995) True bugs of the world (Hemiptera: Heteroptera) classification and natural history. Cornell University Press, Ithaca

Sedlacek JD, Townsend LH (1988) Impact of *Euschistus servus* and *E. variolarius* (Heteroptera: Pentatomidae) feeding on early growth stages of corn. J Econ Entomol 81:840–844

Seiter NJ, Greene JK, Reay-Jones FPF (2013) Reduction of soybean yield components by *Megacopta cribraria* (Hemiptera: Plataspidae). J Econ Entomol 106:1676–1683

Serteyn L, Ponnet L, Backus EA, Francis F (2020a) Characterization of electropenetrography waveforms for the invasive heteropteran pest, *Halyomorpha halys*, on *Vicia faba* leaves. Arthropod Plant Interact 14:113–126

Serteyn L, Ponnet L, Salve M, Fauconnier M-L, Francis F (2020b) Changes of feeding behavior and salivary proteome of brown marmorated stink bug when exposed to insect-induced plant defenses. Arthropod Plant Interact 14:101–112

Silva DP, Fernandes JAM (2016) New evidences supporting trophobiosis between populations of *Edessa rufomarginata* (Heteroptera: Pentatomidae) and *Camponotus* (Hymenoptera: Formicidae) ants. Rev Bras Entomol 60:166–170

Silva FAC, da Silva JJ, Depieri RA, Panizzi AR (2012) Feeding activity, salivary amylase activity, and superficial damage to soybean seed by adult *Edessa meditabunda* (F.) and *Euschistus heros* (F.) (Hemiptera: Pentatomidae). Neotrop Entomol 41:386–390

Spiller NJ, Koenders L, Tjallingii WF (1990) Xylem ingestion by aphids – a strategy for maintaining water balance. Entomol Exp Appl 55:101–104

Steinbauer MJ, Taylor GS, Madden JL (1997) Comparison of damage to *Eucalyptus* caused by *Amorbus obscuricornis* and *Gelonus tasmanicus*. Entomol Exp Appl 82:175–180

Stubbins FL, Mitchell PL, Turnbull MW, Reay-Jones FPF, Greene JK (2017) Mouthpart morphology and feeding behavior of the invasive kudzu bug, *Megacopta cribraria* (Hemiptera: Plataspidae). Invertebr Biol 136:309–320

Sweet MH (1960) The seed bugs: a contribution to the feeding habits of the Lygaeidae (Hemiptera: Heteroptera). Ann Entomol Soc Am 53:317–321

Sweet MH (2000) Seed and chinch bugs (Lygaeoidea). In: Schaefer CW, Panizzi AR (eds) Heteroptera of economic importance. CRC Press, Boca Raton, pp 143–264

Taylor GS (1995) A comparison between the salivary physiology of the crusader bug, *Mictis profana* Fabricius (Coreidae) and the green lucerne mirid, *Creontiades dilutus* (Stål). Ph.D. dissertation, University of Adelaide, Waite, South Australia

Tiwari S, Wratten SD (2019) Biology and management of the New Zealand endemic wheat bug, *Nysius huttoni* (Hemiptera: Lygaeidae). J Integr Pest Manag 10:34;1–10

Torres-Acosta RI, Sánchez-Peña SR, Torres-Castillo JA (2017) Feeding by bagrada bug, *Bagrada hilaris*, on *Moringa oleifera* (Brassicales: Moringaceae) in Mexico. Southwest Entomol 42:919–921

Tuelher ES, Backus EA, Cervantes F, Oliveira EE (2020) Quantifying *Lygus lineolaris* stylet probing behavior and associated damage to cotton leaf terminals. J Pest Sci 93:663–677

Waldkircher G, Webb MD, Maschwitz U (2004) Description of a new shieldbug (Hemiptera: Plataspidae) and its close association with a species of ant (Hymenoptera: Formicidae) in Southeast Asia. Tijdschrift voor Entomologie 147:21–28

Wayadande AC (1994) Electronic monitoring of leafhoppers and planthoppers: feeding behavior and application in host-plant resistance studies. In: Ellsbury MM, Backus EA, Ullman DL (eds) History, development, and application of AC electronic insect feeding monitors. Entomological Society of America, Lanham, pp 86–105

Wayadande A, Bruton B, Fletcher J, Pair S, Mitchell F (2005) Retention of cucurbit yellow vine disease bacterium *Serratia marcescens* through transstadial molt of vector *Anasa tristis* (Hemiptera: Coreidae). Ann Entomol Soc Am 98:770–774

Weintraub PG, Beanland LA (2006) Insect vectors of phytoplasmas. Annu Rev Entomol 51:91–111

Weintraub PG, Wilson MR (2010) Control of phytoplasma diseases and vectors. In: Weintraub PG, Jones P (eds) Phytoplasmas: genomes, plant hosts and vectors. CABI, Wallingford, pp 233–249

Wheeler AG Jr (2001) Biology of the plant bugs (Hemiptera: Miridae). Cornell University Press, Ithaca

Williams L III, Tugwell NP (2000) Histological description of tarnished plant bug (Heteroptera: Miridae) feeding on small cotton floral buds. J Entomol Sci 35:187–195

Chapter 4
Electronic Monitoring of the Feeding Behavior of Phytophagous Stink Bugs (Pentatomidae)

Contents

Abstract Stink bugs (Pentatomidae) encompass a great variety of heteropterans, many of them important pest species. Electronic monitoring of feeding behavior of pentatomids using the electrical penetration graph (EPG) technique has been conducted over a relatively short time span (10 years). To date, eight pentatomid species have been studied with EPG: *Diceraeus* (*Dichelops*) *furcatus* (F.); *Diceraeus* (*Dichelops*) *melacanthus* (Dallas); *Edessa meditabunda* (F.); *Euschistus heros* (F.); *Halyomorpha halys* (Stål); *Nezara viridula* (L.); *Piezodorus guildinii* (Westwood); and *Tibraca limbativentris* Stål. Feeding behavior of adults, and less frequently nymphs, was evaluated on different crop plants but primarily on soybean. These studies are revealing a variety of waveforms. Each one correlates with a specific behavior or a feeding site, which includes holding still on the plant surface, moving, grooming, labial dabbing, stylet penetration and secretion of gelling saliva, formation of a salivary sheath, ingestion from vascular tissues, laceration/maceration of cells, and ingestion of externally digested cell contents. Waveforms recorded in each feeding activity/tissue, are, in general, similar among species studied. Xylem waveforms share appearance and electrical characteristics; phloem ingestion and transitional waveforms (X waves), recorded from only two stink bug species, look alike, as well as waveforms of cell rupture activities observed on parenchyma/

A. R. Panizzi et al., *Electronic Monitoring of Feeding Behavior of Phytophagous True Bugs (Heteroptera)*, Entomology in Focus 6,
https://doi.org/10.1007/978-3-030-64674-5_4

mesophyll and reserve tissue (seeds), independent of the tissue. These generated waveforms, related to their various feeding sites, allow a better understanding of the feeding process of pentatomids and the resulting damage to plants.

Keywords Pentatomidae · Stink bug waveforms · Electronic monitoring of stink bug feeding

4.1 Introduction

In this chapter we will present and discuss electronic monitoring of the feeding behavior of pentatomid species recorded to date using the electrical penetration graph (EPG) technique. As readers will note, this research has been conducted over a relatively short time span. The first heteropteran EPG research goes back less than 30 years, to the publication by Bonjour et al. (1991) of their pioneering EPG study of the coreid, *Anasa tristis* (De Geer), on cucurbit plants.

For several years following the coreid paper, few other studies were carried out. They were spaced several years apart, indicating that few researchers were pursuing the use of EPG to analyze heteropteran feeding. Furthermore, it is worth noting that all these studies used insects (i.e., mirids and early instars of coreids) comparable in size to the aphids and leafhoppers traditionally wired for EPG. Studies with much larger adult stink bugs (Pentatomidae) were not published until recently (Lucini and Panizzi 2016), although interest in conducting EPG with large bugs had begun earlier in that decade with two masters' theses, on *A. tristis* and *Nezara viridula* (L.) late instar nymphs (Maskey 2010; Cooke 2014). In contrast, numerous studies were published over the past 5 years (2016–2020) using EPG to monitor the feeding behavior of stink bugs and other true bug species. Researchers began to appreciate the potential use of EPG, initially devoted to small-soft bodied insects, to study larger and more robust sucking insects.

The Pentatomidae (stink bugs) is the family best represented by far among EPG studies of Heteroptera. In this chapter, therefore, we will focus on the various waveforms generated via EPG for each pentatomid species investigated and summarize their feeding sites and the behaviors with which the waveforms have been correlated. Finally, we will compare the waveforms produced by all stink bugs during their respective feeding activities, highlighting their similarities and differences. In a subsequent chapter, we will consider the more limited EPG research that has been conducted with other heteropteran families and compare waveforms and feeding behaviors across the suborder.

4.2 Research on Pentatomidae Using Electronic Monitoring Methods

Eight pentatomid species have been studied with EPG thus far: *Diceraeus (Dichelops) furcatus* (F.); *Diceraeus (Dichelops) melacanthus* (Dallas); *Edessa meditabunda* (F.); *Euschistus heros* (F.); *Halyomorpha halys* (Stål); *Nezara viridula* (L.); *Piezodorus guildinii* (Westwood); and *Tibraca limbativentris* Stål. Feeding behavior of adults, and less frequently nymphs, was evaluated on different crop plants, but most often on soybean. *E. meditabunda* was the first published pentatomid species evaluated via EPG (Lucini and Panizzi 2016). As follows, we briefly discuss the feeding behaviors performed by these stink bug species and their respective EPG waveforms. In general, the recordings were made using multiple input impedance levels; however, herein we have focused on describing waveform appearance and details registered at 10^7 Ohms, because it is optimal for pentatomids (Lucini and Panizzi 2018a).

4.2.1 Diceraeus (Dichelops) furcatus

Species of the genus *Diceraeus* (= *Dichelops*) are called green-belly stink bugs in South America. Among them, *D. furcatus* is a Neotropical species more adapted to subtropical areas (Panizzi 2015), where it has been reported as a secondary pest on soybean since the 1970s (Panizzi et al. 1977). More recently, this stink bug has been observed feeding on several species of spring cereals at different plant developmental stages (mostly seedlings and immature seed heads) (Panizzi et al. 2018). Wheat plants are frequently damaged during early plant development (tillering) and booting/milk-grain stages in Southern Brazil. This damage includes tissue necrosis and deformation of leaves and undeveloped and discolored seed heads (symptom known as "white seed head") (Panizzi et al. 2016).

The feeding behavior of *D. furcatus* adult females was recorded on vegetative (stem at V3 – tillering; Large [1954]) and reproductive (seed head at R11.1 – milk grain) developmental stages of wheat plants using a four-channel AC-DC monitor (Backus et al. 2019; EPG Technologies, Inc., Gainesville, FL, USA). Nine different waveform types/subtypes were identified, where two waveforms represent non-feeding activities (Np and Z) and seven represent feeding activities (Df1a, Df1b, Df2, Df3a, Df3b, Df4a, and Df4b). These waveforms and their proposed biological meanings are summarized in Table 4.1.

Non-feeding activity is represented by two clearly distinguishable waveforms (Z and Np). Waveform Z is related to the bug resting on the plant surface; it represents the baseline of the recording and appears as a flat waveform with very slight variation in the voltage level. The second non-feeding waveform (Np) is observed when the stink bug is moving on the plant surface; it is displayed as a highly irregular waveform (i.e., without a clear pattern) containing multiple peaks randomly

Table 4.1 Summary of the EPG waveforms of *Diceraeus* (*Dichelops*) *furcatus* adults recorded on stem (V3 stage – tillering) and seed head (R11.1 stage: milk grain) of wheat plants and their proposed biological activities

Phase	Family	Type or subtype	Plant structure observed	Proposed biological activities
Non-feeding	–	Z	Stem/seed head	Standing still on the plant surface
		Np	Stem/seed head	Walking on the plant surface
Pathway	P	Df1a	Stem/seed head	Stylet penetration and salivary sheath secretion
		Df1b	Seed head	Bug encountering a rigid cell layer requiring stylet protraction and retraction
Ingestion	I	Df2	Stem/seed head	Xylem sap ingestion
Salivation/ingestion	I	Df3a	Stem	Cell laceration, enzymatic maceration of stem tissues and probably ingestion
Ingestion	I	Df3b	Stem	Short ingestion of macerated stem tissues
Salivation	I	Df4a	Seed head	Cell laceration, enzymatic maceration of seed endosperm
Ingestion	I	Df4b	Seed head	Ingestion of macerated seed endosperm

Source: adapted from Lucini and Panizzi (2017b)

distributed. These irregular signals are caused by electrical contacts of tarsal claws while scratching the plant surface (Lucini and Panizzi 2017b) (see Chap. 8 for illustration and further discussion of waveforms of non-feeding behaviors).

The pathway phase begins when the bug inserts its stylets into plant tissue to probe it for acceptance or rejection of the feeding site, i.e., this phase represents the first electrical contact between the mouthparts (stylets) and plant tissue. As the bug penetrates the electrified plant tissue, a sudden increase or decrease of the voltage level is observed in the recording; the direction depends on the Ri level applied. The pathway phase of *D. furcatus* consists of a single waveform type (named Df1), which is divided into two subtypes: Df1a and Df1b. The first one has been recorded on both stems and seed heads of wheat, and it presents an irregular pattern with several peaks. In contrast, the subtype Df1b is observed only on wheat seed heads and is marked by a more regular waveform appearance (Lucini and Panizzi 2017b).

Based on histological studies, Df1a represents the initial stylet insertion, secretion of gelling saliva to create a salivary sheath, and deeper stylet penetration into plant tissue toward the feeding site. The waveform Df1b is probably related to presence of a rigid cell layer that the bug faced during stylet penetration which it needed to overcome to reach the feeding site, primarily seed endosperm (Lucini and Panizzi 2017b). The salivary sheath deposited in the parenchyma tissue during pathway phase is either completely or incompletely secreted, depending on the feeding site. A complete salivary sheath is created when Df1 precedes xylem ingestion (waveform Df2), whereas an incomplete salivary sheath is secreted when Df1 is recorded before cell-rupture feeding on stems and seed heads (waveforms Df3 and Df4, respectively) (Lucini and Panizzi 2017b). The secretion of a complete or incomplete salivary sheath depends on the feeding strategy used by the bugs (see Chap. 2 for further discussion).

Following the Df1 waveform, the ingestion phase is recorded, which comprises five waveforms, Df2, Df3a, Df3b, Df4a, and Df4b, each related to a specific feeding tissue. The waveform Df2 occurs on both vegetative (stem) and reproductive (seed head) developmental stages. Df2 is extremely similar between the two plant structures; it shows a very regular pattern composed of wave portions interspersed with peaks regularly distributed over time (see Fig. 4.4, Sect. 4.3). Df2 is correlated, via histological observations, with xylem sap ingestion where the bugs use the salivary sheath strategy (Lucini and Panizzi 2017b).

Waveform Df3 recorded on wheat stem is divided into subtypes Df3a and Df3b, which occur interspersed during the entire waveform recording. Df3a is primarily an irregular waveform with several peaks and greatly variable in appearance. During visual observations of the waveform Df3a, the stylets were detected moving deeply and continuously inside the stem tissue; this suggests that *D. furcatus* also uses the cell rupture feeding strategy while feeding on wheat stems, employing the lacerate-and-flush tactic and probably maceration, as well (Lucini and Panizzi 2017b). Subtype Df3b is a shorter and more regular waveform compared to Df3a. During this waveform, visual observations demonstrate that stylets stay briefly motionless (for a few seconds) inside the stem tissue. This probably represents short periods of ingestion of the cell contents previously degraded by rupturing activities (Lucini and Panizzi 2017b) (see Fig. 4.6a, Sect. 4.3, for illustration).

Waveform Df4 is recorded on wheat seed head (seed endosperm) only, and it is divided into subtypes, Df4a and Df4b, that occur interspersed. Waveform Df4a primarily shows an irregular appearance (sometimes with regular portions) with several peaks. During Df4a, *D. furcatus* quickly moves its stylets back and forth, deep in the seeds, suggesting cell rupture strategy (laceration and maceration tactics) on wheat seed heads (see Fig. 4.6b and c, Sect. 4.3, for illustration). Histological sections indicated stylet tips positioned in the seed endosperm during Df4a, and an incomplete salivary sheath deposited internally (Lucini and Panizzi 2017b). Df4b is a short waveform, highly regular, resembling Df3b on stems. During Df4b stylets remain motionless inside the plant tissue (in this case in the seed) for a short period of time, indicating ingestion of the cell contents (Lucini and Panizzi 2017b). Severe tissue damage of stems and immature seeds is apparent after the occurrence of Df3 and Df4 waveforms, respectively (Lucini and Panizzi 2017b).

4.2.2 Diceraeus (Dichelops) melacanthus

The green-belly stink bug *D. melacanthus* was rarely found in the neotropics in the 1970s and 1980s (Panizzi and Slansky 1985). Nowadays, this species is widely distributed in the Neotropical Region as a pest; in Brazil, populations increased substantially on soybean, maize, and wheat crops (Panizzi 2015). On soybean plants, bugs prefer to feed on pods (seeds), whereas on maize they feed on stems in earlier stages of plant development (seedlings). The attack on maize seedlings causes tissue necrosis, deformation of leaves, and unproductive tillering.

EPG recordings of *D. melacanthus* adult females feeding on stems of maize seedlings (at V2-V3 stage, 2-3 leaves completely developed, respectively [Ritchie and Hanway 1989]) were made using a four-channel AC-DC monitor. Seven different waveforms were identified, where three represent non-feeding (Np, Z, and O) and four represent feeding activities (Dm1, Dm2, Dm3a, and Dm3b) (Lucini and Panizzi 2017a). These waveforms and their proposed biological meanings are summarized in Table 4.2.

The non-feeding waveform Z represents the bug standing still on the plant surface and is characterized by a flat waveform. Waveform Np has an irregular appearance with several peaks; the amplitude varies with the Ri level applied. It is correlated with movement on the plant surface. Waveform O is marked by a single, large, flat plateau; each plateau was visually correlated with the deposition of one egg on the maize stem surface (Lucini and Panizzi 2017a) (see Chap. 8 for illustrations).

After the stink bug inserts its stylets into maize stem tissue, a single pathway waveform is recorded, Dm1, with a pattern of irregular peaks. Based on histological analysis, Lucini and Panizzi (2017a) proposed that Dm1 represents initial and deep stylet penetration into stem tissue toward the feeding site, accompanied by secretion of a salivary sheath. Bugs move their stylets toward xylem or stem parenchyma; stylet activity in those tissues is represented by waveforms Dm2 and Dm3, respectively. Dm2 is composed of a regular pattern of repetitive wave portions interspersed

Table 4.2 Summary of the EPG waveforms of *Diceraeus* (*Dichelops*) *melacanthus* adults recorded on maize seedlings (stem at V2 – V3 stage) and their proposed biological activities

Phase	Family	Type or subtype	Proposed biological activities
Non-feeding		Z	No movement on the stem surface
		Np	Walking on the stem surface
		O	Egg laying by females on stem
Pathway	P	Dm1	Stylet penetration and salivary sheath secretion
Ingestion	I	Dm2	Xylem sap ingestion
Salivation/ingestion	I	Dm3a	Cell laceration, enzymatic maceration of stem tissues and ingestion
Ingestion	I	Dm3b	Short ingestion period

Source: adapted from Lucini and Panizzi (2017a)

with peaks regularly displayed (Lucini and Panizzi 2017a) (see Fig. 4.4, Sect. 4.3, for illustration). Dm3 is divided into subtypes Dm3a and Dm3b, which occur interspersed. Dm3a is primarily irregular with peaks randomly distributed, and it does not resemble any waveform related to ingestion (Lucini and Panizzi 2017a).

Dm3a is related to cell rupture strategy (laceration and maceration tactics), with the bugs moving their stylets back and forth continuously, deep within the maize stem. Dm3b is regular and short compared to Dm3a. During Dm3b stylets briefly remain static inside stem tissue (for a few seconds), which represents ingestion of cell contents previously degraded via lacerate/macerate activities (Lucini and Panizzi 2017a).

4.2.3 Edessa meditabunda

The brown-winged stink bug, *E. meditabunda*, is a Neotropical pentatomid reported as a pest of plants in Solanaceae and Fabaceae (references in Smaniotto and Panizzi 2015). In Brazil, this species is abundant on soybean, particularly in the central west region. *E. meditabunda* has a preference for vegetative structures (Silva et al. 2012), differing from most other stink bugs which prefer seeds and fruits.

The first published pentatomid study of EPG was with *E. meditabunda* (Lucini and Panizzi 2016), using a Giga-8 model DC-monitor (EPG Systems, Wageningen, The Netherlands) (Tjallingii 1988). The EPG waveforms were obtained for adult females feeding on stems of soybean seedlings (V1 stage – unifoliolate nodes [Fehr et al. 1971]). Eight waveform types were recorded; two represent non-feeding activities (Np and Z), and six represent feeding activities (X, Em1, Em2, Em3, Em4, and Em5). These waveforms and their proposed biological meanings are summarized in Table 4.3.

Waveform Z and Np represent non-feeding activities; Z represents the bug resting (static) on the stem surface and appears as a flat line with slight or no variation in appearance. Np presents multiple irregular peaks, highly variable in amplitude, and includes two behaviors: walking on the plant surface and touching the rostrum (labium) to the plant surface before stylet insertion (Lucini and Panizzi 2016).

Em1 represents stylet insertion into plant tissue. It has an irregular pattern and correlates with initial and deep stylet penetration plus secretion of gelling saliva. This creates a salivary sheath that completely surrounds the stylets as they penetrate the plant tissue until reaching the vascular tissues (salivary sheath strategy). Subsequently, the bug ingests from either the xylem (Em2) or the phloem (Em3) (Lucini and Panizzi 2016).

Em2 is characterized by regular wave portions interspersed with upward peaks, regularly distributed over time (see Fig. 4.4, Sect. 4.3). Em3 presents an overall stereotypical sinuous pattern, alternating up and down, with a fine structure of very small, highly regular waveforms composing this sinuous wave (see Fig. 4.5, Sect. 4.3). As the stylets were severed during Em3, exudation of droplets (probably sap)

Table 4.3 Summary of the EPG waveforms of *Edessa meditabunda* adults recorded on soybean stems (V1 stage) and their proposed biological activities

Phase	Family	Type	Proposed biological activities
Non-feeding	–	Z	Standing still
		Np	Walking on the plant surface and labial tapping
Pathway		Em1	Stylet pathway/salivary sheath secretion
X wave		X	Penetration of phloem cell; salivation, egestion, tasting of the cell
Ingestion	I	Em2	Xylem sap ingestion
		Em3	Phloem sap ingestion
Interruption	N	Em4	Salivary interruptions during xylem ingestion
		Em5	Salivary interruptions during phloem ingestion

Source: adapted from Lucini and Panizzi (2016)

from the plant was observed, likely due to the high internal hydrostatic pressure of the phloem sieve tube elements (Taiz and Zeiger 2004).

A transitional waveform, named the X wave, is always observed preceding waveform Em3. This X wave is marked by a highly regular waveform with an abrupt voltage drop as it finishes and the Em3 waveform begins (see Fig. 4.3, Sect. 4.3). X waves are transitional, multi-component waveforms that consist of several stylet activities in the preferred ingestion tissue (e.g., watery salivation and egestion) to taste/test the cells for recognition. X waves are primarily reported for xylem and phloem sap feeders preceding sustained ingestion (Backus et al. 2009). For *E. meditabunda*, the X wave was similarly observed prior to phloem activities and probably represents penetration into a sieve element, salivation, and tasting/testing for acceptance, followed by Em3, ingestion of cell contents (Lucini and Panizzi 2016).

During feeding in vascular tissues, short interruption waveforms, Em4 and Em5, were observed during xylem and phloem ingestion, respectively. Em4 is irregular and recorded near the start of xylem ingestion (Lucini and Panizzi 2016). Interruptions during xylem ingestion in sharpshooters represent watery salivation and/or putative tasting/testing of cells (Backus et al. 2005a).

Em5 is also irregular; however, it may be observed throughout the entire course of phloem ingestion (Lucini and Panizzi 2016). In aphids, salivation behavior is observed at the beginning of the phloem phase and is represented by a sole waveform (E1) (Prado and Tjallingii 1994). In *E. meditabunda*, Em5 probably represents the only salivation during phloem activities and occurs interspersed with ingestion events (Lucini and Panizzi 2016).

4.2.4 Euschistus heros

The Neotropical brown stink bug *E. heros* was rarely found in the Neotropics in the 1970s (Panizzi et al. 1977); presently, it is the most important pest damaging soybean (Panizzi et al. 2012). It is reported as a potential invasive species in the USA (Panizzi 2015). *E. heros* feeds on several plant species from different plant families,

Table 4.4 Summary of the EPG waveforms of *Euschistus heros* adults recorded on soybean pod (R5 stage – pod filling) and their proposed biological activities

Phase	Family	Type or subtype	Proposed biological activities
Non-feeding		Z	Standing still on the plant surface
		Np	Walking on the plant surface
		Dw1	Egestion of saliva/regurgitation of liquid food on pod surface
		Dw2	Probably re-ingestion of saliva/regurgitation of liquid food
Pathway	P	Eh1a	Beginning of stylet penetration and secretion of gelling saliva to form a salivary sheath
		Eh1b	Deep stylet penetration and secretion of gelling saliva to form branches of a salivary sheath
		Eh1c	Bug encountering a rigid cell layer requiring stylet protraction and retraction
		Eh1w	Stylet withdrawal from the plant tissue
Ingestion	I	Eh2	Sustained xylem sap ingestion
Salivation	I	Eh3a	Cell laceration and enzymatic maceration of endosperm tissue
Ingestion	I	Eh3b	Short ingestion event of lacerated/ macerated endosperm tissue
	I	Eh4	Short ingestion event from an unknown site
Interruption	N	Eh5	Short interruptions during xylem sap ingestion

Source: adapted from Lucini and Panizzi (2018b)

but it prefers plants in the family Fabaceae (legumes) (Smaniotto and Panizzi 2015), with a strong preference for soybean, where they feed primarily on seeds.

EPG waveforms were recorded for *E. heros* adult females feeding on soybean pods (at R5 stage – pod filling) using a four-channel AC-DC monitor. Thirteen different waveform types/subtypes were generated, of which four represent non-feeding (Z, Np, Dw1, and Dw2), and nine represent feeding activities (Eh1a, Eh1b, Eh1c, Eh1w, Eh2, Eh3a, Eh3b, Eh4, and Eh5). These waveforms and their proposed biological meanings are summarized in Table 4.4.

The non-feeding waveform Z occurs while the bug is resting on the pod surface and displays with a flat waveform without changes in appearance. Np, with an irregular appearance and multiple peaks, represents the stink bug walking on the pod surface. Dw1 and Dw2 represent a curious behavior not yet reported in EPG studies. Dw1 is irregular with large plateaus containing sporadic peaks; Dw2 is a regular waveform, similar to ingestion-related waveforms (Lucini and Panizzi 2018b).

Dw1 is correlated with excreted droplets of liquids being deposited/dragged on the pod surface using the tip of the rostrum (labium). This waveform represents egestion of liquids, which may contain saliva and/or regurgitated liquid ingested from xylem, since this activity is primarily recorded after occurrence of long events of xylem sap ingestion. During Dw2, the tip of the rostrum is inside the liquid droplet deposited on the pod surface. Dw2 represents the re-ingestion of small amounts

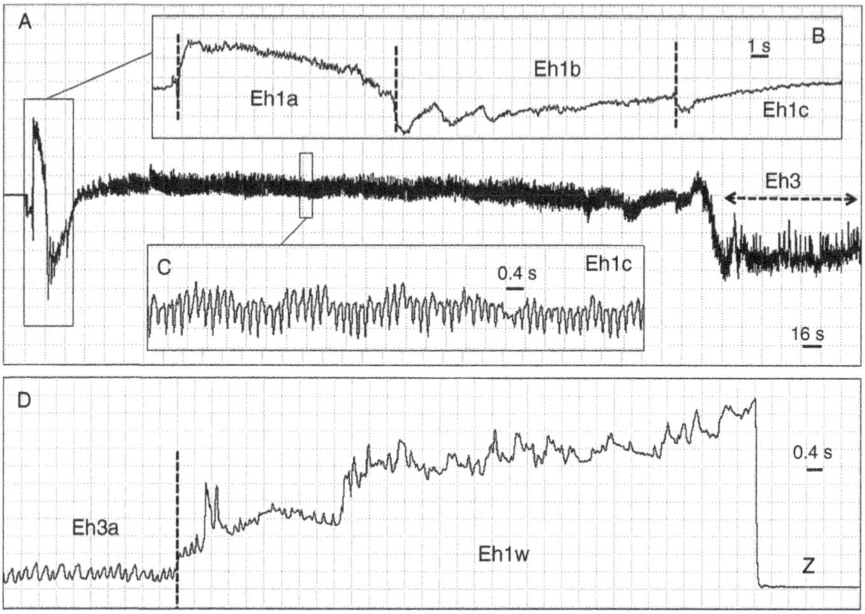

Fig. 4.1 Waveforms recorded from *Euschistus heros* on a soybean pod (50 mV AC applied voltage and input impedance of 10^7 Ohms). Early feeding activity showing the pathway waveforms (Eh1) preceding the cell rupture waveform (Eh3) (**a**). Expanded views of the pathway subtypes Eh1a (initial stylet insertion), Eh1b (deep stylet penetration), and Eh1c (stylets faced with a physical barrier – sclerenchyma) (**b** and **c**). Expanded view of the pathway waveform Eh1w (stylet removal), recorded between waveforms Eh3 and Z (non-feeding activity) (**d**). (Source: adapted from Lucini and Panizzi (2018b))

of liquid previously secreted (Lucini and Panizzi 2018b) (see Chap. 8 for illustration and further discussion).

Unlike that of the majority of pentatomids evaluated, the pathway phase of *E. heros* comprises multiple waveforms: Eh1a, Eh1b, Eh1c, and Eh1w. The first three subtypes represent contact of the rostrum with plant tissue and the initial and deep stylet penetration into pod tissue. Eh1w is correlated with stylet removal after feeding activities (Fig. 4.1). Eh1a is short, composed of irregular peaks; it represents the initial stylet penetration into the pod and secretion of salivary sheath internally (Lucini and Panizzi 2018b).

Eh1b is characterized as flat waveform, with irregular appearance. It represents deeper stylet penetration into pod tissue (parenchyma cells) and secretion of gelling saliva to create salivary sheath branches (Lucini and Panizzi 2018b). Eh1c is a stereotypical waveform with longer duration and a more regular pattern compared to other pathway waveforms. Eh1c is primarily recorded preceding seed endosperm activities (waveform Eh3 – see below) and rarely before xylem feeding (waveform Eh2). During Eh1c the bug repeatedly forces the head downwards, pushing the stylets deeper inside the pod tissue, and then pulls them upward (Lucini and Panizzi 2018b). These head movements suggest the presence of a physical barrier that bugs need to overcome before reaching the seed endosperm. Histological sections made during Eh1c show stylet tips anchored in the sclerenchyma, a rigid cell layer (Lucini and Panizzi 2018b). The last pathway subtype, Eh1w, is a short waveform with irregular peaks always recorded when the bug removes its stylets from the pod tissue at the end of a feeding activity, returning to non-feeding waveforms (Z or Np) (Lucini and Panizzi 2018b).

During the ingestion phase, *E. heros* exploit two feeding sites (represented by waveforms Eh2 and Eh3) and one still unknown site (represented by waveform Eh4) (Lucini and Panizzi 2018b). Eh2 is a stereotypical waveform with regular wave portions interspersed with regularly distributed peaks (see Fig. 4.4, Sect. 4.3). Based on histological sections and similarities with other pentatomid waveforms, Eh2 is correlated with xylem sap ingestion. Sometimes during xylem ingestion (mostly at the beginning), a short interruption waveform is observed (interruption phase – waveform Eh5). Eh5 is composed of a flat-spiky plateau with an irregular pattern (Lucini and Panizzi 2018b). This interruption waveform probably represents quick secretion of watery saliva to taste/test the xylem cells (Backus et al. 2013).

The second ingestion waveform (Eh3) represents feeding activity in the seeds; it is clearly separated into two subtypes, Eh3a and Eh3b, which occur interspersed over the entire waveform recording. Eh3a is composed of irregular peaks with no clear pattern. Visual observations during Eh3a show bugs moving their stylets in and out quickly and continuously within the pod tissue (i.e., seed endosperm, which was demonstrated in the histological sections). In addition, cuts made in fresh soybean pods, after an Eh3 event, show a huge damaged area in the seeds (Lucini and Panizzi 2018b). This indicates that *E. heros* uses the cell rupture strategy with lacerating and macerating tactics to degrade seed tissue before ingesting cell contents (see Fig. 4.6b and c). Eh3b is a highly regular waveform without peaks and with very short duration compared to Eh3a. During Eh3b, stylets are briefly static inside the pod tissue, indicating ingestion of seed contents previously degraded (Lucini and Panizzi 2018b).

Eh4 was not often observed during recordings. It shows a regular pattern with a short duration (<60 seconds). Stylet tips were observed near vascular tissues during Eh4 (Lucini and Panizzi (2018b). Eh4 shares similarities with waveform Em3 from *E. meditabunda*, which is histologically correlated with phloem sap ingestion (Lucini and Panizzi 2016), but the biological meaning of waveform Eh4 remains uncertain.

4.2.5 Halyomorpha halys

The brown marmorated stink bug, *H. halys*, is a species native to Asia. It is highly polyphagous, feeding on over 175 plant species including cultivated and non-cultivated hosts, and is widely spread around the world. A strong preference for fruits and seeds is evident, but other plant structures are also attacked (Nielsen and Hamilton 2009; Hamilton et al. 2018). Besides agricultural damage, *H. halys* causes

Table 4.5 Summary of the EPG waveforms of *Halyomorpha halys* nymphs recorded on faba bean leaves (2-weeks-old) and their proposed biological activities

Phase	Family	Type or subtype	Proposed biological activities
Non-feeding	–	Z/Np	Standing still/walking on the plant surface
Epidermal scanning	–	Hh1	Sensory evaluation by scanning movements of labium; slight breakage of plant cuticle
Pathway/cell rupturing	P	Hh2	Stylet penetration with salivary sheath secretion or cell rupturing without salivary sheath secretion
X wave	P/I	Hh3a-1	Xylem activities: salivation/egestion
		Hh3a-2	Xylem activities: trial ingestion
		Hh3b	Phloem activities: penetration, watery salivation/egestion, tasting
Ingestion	I	Hh4a	Xylem sap ingestion
		Hh4b	Putative phloem sap ingestion
		Hh5	Putative macerate ingestion

Source: adapted from Serteyn et al. (2020)

nuisance problems during winter, when a large number of individuals invade human-made buildings to overwinter (Inkley 2012).

The first electronic monitoring of feeding behavior of *H. halys* was performed using a modified EMIF system, which does not produce waveforms (see Chap. 1 for more details) (Wiman et al. 2014). Recently, EPG recordings of *H. halys* fifth instars feeding on faba bean leaves were made using a Giga-8 DC monitor and yielded 10 different waveform types/subtypes: Np, Z, Hh1, Hh2, Hh3a-1, Hh3a-2, Hh3b, Hh4a, Hh4b, and Hh5 (Serteyn et al. 2020). These waveforms and their proposed biological meanings are summarized in Table 4.5.

Two waveform types in the non-feeding activities include Z and Np, which occur interspersed. Z is a flat waveform, and Np is represented by events with high and irregular peaks. Z represents the insect standing still on the leaf surface (i.e., the baseline), whereas Np represents the insect walking on the leaf surface or touching it with its rostrum (labium) (Serteyn et al. 2020).

Before pathway, the bug performs epidermal scanning, i.e., sensory evaluation of the plant surface with the tip of the labium, probably involving a slight breakage of the plant cuticle. This waveform, Hh1, is characterized by a sudden voltage drop, followed by irregular peaks and then a slow return to slightly above the baseline (Serteyn et al. 2020). Simple contact between the labium and the plant surface is known as "labial dabbing" and has been observed in many hemipterans. Labial dabbing may involve secretion of watery saliva for precibarial tasting (Backus 1988) but has not been reported to involve scraping of the plant epidermis as also proposed for the Hh1 waveform in *H. halys* (Serteyn et al. 2020). *H. halys* often performed epidermal scanning while standing still on the plant surface, and this behavior, considered distinct from labial dabbing, is hypothesized to facilitate selection of a penetration site (Serteyn et al. 2020).

After scanning the plant surface, the bug inserts its stylets in the tissue and penetrates deeply, producing waveform Hh2 that represents the pathway/cell rupturing phase. Hh2 is repeated several times per insect during the recordings. Some Hh2 events are short, with irregular peaks, and do not precede any other feeding waveform, i.e., the bug removes the stylets from the plant following Hh2. This behavior is interpreted to represent test probes, which are not followed by ingestion. Sometimes Hh2 shows a longer duration, becoming a little more regular, with peaks regularly distributed, and precedes other feeding waveforms (Serteyn et al. 2020).

H. halys probably secretes gelling saliva to form a salivary sheath during the pathway phase, although histological evidence is lacking. However, EPG studies with other true bugs ingesting from vascular tissues demonstrate that a salivary sheath is secreted during pathway, and the salivary sheath strategy is used (e.g., Backus et al. 2013; Lucini and Panizzi 2016; Mitchell et al. 2018). A resemblance between Hh2 and the waveform correlated with cell rupture in *Lygus* spp. (Cervantes et al. 2016) suggested that this pathway waveform might also be associated with cell rupturing by *H. halys*.

Preceding some ingestion waveforms, *H. halys* performs transition behaviors (Hh3a and Hh3b) that represent the X wave phase. *H. halys* is the first heteropteran species to perform two distinct X waves, each one preceding ingestion from a different tissue; this has also been reported for the leafhopper, *Scaphoideus titanus* Ball (Chuche et al. 2017).

Hh3a is composed of repetitive plateaus (episodes) with small spikes. Each episode is divided into subtypes Hh3a-1 and Hh3a-2, which differ mostly by peak orientation. Hh3a precedes waveform Hh4a, which represents sustained xylem ingestion; thus, it is considered a xylem X wave. Hh3b displays an irregular appearance, similar to Hh2. It is recorded preceding the waveform Hh4b, which is hypothesized to represent ingestion from phloem sieve elements; thus, Hh3b is considered a putative phloem X wave (Serteyn et al. 2020).

The ingestion phase comprises waveforms Hh4a, Hh4b, and Hh5. Waveforms Hh4a and Hh4b represent about 50% and 25% of the total ingestion events performed by *H. halys* on faba bean leaves, respectively. These two waveforms are similar; both contain peaks regularly distributed between the wave portions. They differ by peak orientation and relative amplitude (Serteyn et al. 2020). Histological sectioning during Hh4 indicates xylem sap ingestion; furthermore, Hh4a is similar to xylem waveforms recorded from other heteropterans (e.g., Backus et al. 2013; Lucini and Panizzi 2018a). Serteyn et al. (2020) correlated Hh4b with putative phloem sap ingestion; however, no definitive histological evidence was obtained. This hypothesis is based on similarity in appearance and electrical characteristics between the X wave (Hh3b) of *H. halys* with the phloem X wave recorded for chinch bugs.

The last ingestion waveform (Hh5) is not preceded by an X wave and represents the remaining 25% of the total ingestion events performed by *H. halys* on faba bean leaves. This waveform is marked by an almost-regular pattern with short plateaus, but without presence of peaks, as observed for Hh4 waveforms. Serteyn et al. (2020) propose that Hh5 might represent salivation/maceration of mesophyll/parenchyma tissues and, simultaneously, ingestion of the degraded cell contents, i.e., cell rupturing activities. Hh5 resembles waveform Dm3a of the pentatomid *D. melacanthus* (Lucini and Panizzi 2017a), which is correlated with cell rupturing feeding activity on maize seedlings.

4.2.6 Nezara viridula

The southern green stink bug or green vegetable bug, *N. viridula*, is a pentatomid species widely distributed in almost all zoogeographical regions of the world (Todd 1989). It is extremely polyphagous and for a long time was considered the major stink bug species damaging several plant crops in the Americas, including soybean, cotton, and vegetables (McPherson and McPherson 2000; Panizzi and Lucini 2016). In the last 20 years, *N. viridula* population declined substantially in the Americas, becoming, in general, a secondary pest (Panizzi and Lucini 2016); however, it is still considered an important stink bug pest elsewhere in the world (Esquivel et al. 2018). *N. viridula* prefers to feed on reproductive structures (fruits and immature seeds), but it is also reported to feed on vegetative structures of cultivated and non-cultivated host plants (McPherson and McPherson 2000; Panizzi 1997; Smaniotto and Panizzi 2015).

EPG waveforms for *N. viridula* fifth instars feeding on petiole (V3-V6 stage; 2nd–5th trifoliate leaves) and immature pod (R5 stage – pod filling) were recorded on an early version of the 4-channel AC-DC monitor (Backus and Bennett 2009) and generated seven waveform types/subtypes: three represent non-feeding activities (Z1, Z2, and Z3) and four represent feeding activities (Nv1a, Nv1b, Nv2, and Nv3) (Mitchell et al. 2018). These waveforms and their proposed biological meanings are summarized in Table 4.6.

The non-feeding phase comprises three distinct waveforms Z1, Z2, and Z3. Z1 is a flat line (baseline of the recording) and is correlated with bugs standing still on the plant surface. Z2 shows irregular peaks and is correlated with bugs walking or grooming. Z3 also presents irregular peaks (as sharp spikes), but with a large variation in the amplitude; it represents labial dabbing and often precedes the initial stylet insertion into plant tissue (Mitchell et al. 2018).

Pathway phase is represented by waveform Nv1, which is divided into two subtypes Nv1a and Nv1b. Nv1a is the first pathway waveform recorded. It is marked by irregular high spikes. Nv1b is also irregular, but characterized by recognizable episodes, each one composed of humps that decrease in voltage level over time. These two pathway waveforms represent the initial stylet penetration into plant tissue toward the feeding site and secretion of a salivary sheath either externally or internally in the plant tissue (Mitchell et al. 2018).

The ingestion phase comprises two distinct waveforms, Nv2 and Nv3, which represent ingestion activities from different tissues. Nv2 was recorded in all plant structures evaluated (petiole, stem, and pod) and presents similar characteristics and

Table 4.6 Summary of the EPG waveforms of *Nezara viridula* fifth instars recorded on vegetative (petiole at V3-V6 stage) and reproductive (pod at R5 stage) development stages of soybean plants and their proposed biological activities

Phase	Family	Type or subtype	Plant structure	Proposed biological activities
Non-feeding	Z	Z1	Petiole/pod	Holding still on plant surface
		Z2	Petiole/pod	Walking, grooming, etc.
		Z3	Petiole/pod	Labial dabbing
Pathway	P	Nv1a	Petiole/pod	Initial stylet penetration and secretion of gelling saliva
		Nv1b	Petiole/pod	Passage of stylets through non-preferred plant tissues and formation of salivary sheath
Ingestion	I	Nv2	Petiole/pod	Ingestion from xylem
Salivation plus ingestion	I	Nv3	Pod	Laceration, production of watery saliva, ingestion of externally digested seed contents

Source: adapted from Mitchell et al. (2018)

appearance among them. This waveform displays a very regular pattern composed by peaks interspersed with wave portions. Histological observations made during Nv2 indicate xylem sap ingestion (Mitchell et al. 2018). Subsequent research (Rivera and Mitchell 2020) showed that first instars ingesting from leaves produced only the Nv2 waveform. In first and fifth instars and adults, Nv2 strongly resembles xylem waveforms recorded for other pentatomids (Lucini and Panizzi 2018a).

The second ingestion waveform, Nv3, only occurs on soybean pods, never on vegetative tissue. It is composed of irregular spikes, highly variable in appearance. Nv3 represents cell rupturing feeding activities on seed endosperm or pod wall tissue (Mitchell et al. 2018). Cell rupture feeding in pentatomids is divided into two subtypes that, in general, occur interspersed. One subtype represents the laceration/maceration tactics, and the other is correlated with ingestion of previously degraded cell contents (Lucini and Panizzi 2018a) (Fig. 4.6b). For *N. viridula*, these two distinct waveforms could not be distinguished (Mitchell et al. 2018), and Nv3 probably represents a combination of cell degradation via cell rupture activities plus ingestion of contents.

At the end of feeding, stylets are pulled out, and at this time, a short and irregular waveform, similar to Nv1a, is recorded. Salivary flanges remaining on the plant after pentatomid feeding are generally sealed, although some remain open (Hollay et al. 1987). Mitchell et al. (2018) suggest that this waveform, during stylet removal, represents the secretion of additional gelling saliva to seal the salivary sheath previously formed.

4.2.7 Piezodorus guildinii

The redbanded stink bug, *P. guildinii*, is a Neotropical stink bug species, widespread in the Americas and a key pest of soybean (Panizzi and Slansky 1985; Bundy et al. 2018). *P. guildinii* feeds on a variety of cultivated and non-cultivated plants but exhibits a great preference for legumes (Fabaceae) (Smaniotto and Panizzi 2015). The first published study of a pentatomid seed feeder using EPG (Lucini et al. 2016) was done with this species.

The EPG for *P. guildinii* adult females feeding on vegetative (leaflet and stem at V2 stage – 1st trifoliate leaves) and reproductive structures (pods at R5 stage – pod filling) of soybean plants, recorded on a 4-channel AC-DC monitor, generated eleven waveform types/subtypes; three represent non-feeding activities (Np, Z, and R), and eight represent feeding activities (Pg1a, Pg1b, Pg1c, Pg1d, Pg2, Pg3a, Pg3b, and Pg4) (Lucini et al. 2016). These waveforms and their proposed biological activities are summarized in Table 4.7.

Behaviors related to non-feeding activities include waveform Z, which is a waveform with no change in appearance over time and represents the stink bug static (resting). The second type, named Np, presents several peaks and an irregular

Table 4.7 Summary of the EPG waveforms of *Piezodorus guildinii* adults recorded on vegetative (leaf and stem at V2 stage) and reproductive (pod at R5 stage) development stages of soybean plants and their proposed biological activities

Phase	Family	Type or subtype	Plant structure	Proposed biological activities
Non-feeding	–	Np	Leaflet/stem/pod	Walking on the plant surface
		Z	Leaflet/stem/pod	Standing still on the plant surface
		R	Leaflet/stem/pod	Tapping plant surface with rostrum (labium)
Pathway	P	Pg1a	Leaflet/stem/pod	Beginning of stylet penetration and secretion of thickest gelling saliva to form the trunk of the salivary sheath
		Pg1b	Leaflet/stem/pod	Stylet penetration and secretion of gelling saliva to form branches of a salivary sheath
		Pg1c	Leaflet/stem/pod	Secretion of thinner gelling saliva and deeper penetration of stylets
		Pg1d	Stem/pod	Stylet protraction and retraction as bug encounters a rigid cell layer
Ingestion	I	Pg2	Leaflet/stem/pod	Xylem sap ingestion
Salivation	I	Pg3a	Pod	Watery salivation into seed endosperm
Ingestion	I	Pg3b	Pod	Ingestion of macerated seed endosperm
Interruptions	N	Pg4	Leaflet/stem/pod	Brief interludes of salivation during xylem ingestion

Source: adapted from Lucini et al. (2016)

appearance; it is visually correlated with movement by the stink bug on the plant surface. Finally, waveform R is recorded, but not frequently as Z and Np. Waveform R is characterized by large and regular peaks and visually correlated with the bug touching the plant surface with the tip of the rostrum (labium) (Lucini et al. 2016).

The pathway phase comprises one waveform type (Pg1) divided into four subtypes: Pg1a, Pg1b, Pg1c, and Pg1d. All these waveforms occur on the three plant tissues tested (leaflet, stem, and pod), except waveform Pg1d, which is observed sporadically on stems and primarily on pods (Lucini et al. 2016). Waveform Pg1a represents the first electrical contact between stylets and the plant tissue. Pg1a has a very short duration (a few seconds) and is marked by irregularly distributed peaks. Pg1b consists of a repetitive series of stereotypical episodes, each one represented by a sudden increase of the voltage level followed by a gradual decrease. Pg1c is mostly represented by a flat waveform with irregular appearance. Based on histological observations and comparison with pathway waveforms from other sucking insects, Pg1a, Pg1b, and Pg1c represent initial and deep stylet penetration plus secretion of gelling saliva to form a salivary sheath (Lucini et al. 2016).

Pg1d displays a very regular appearance compared with the earlier pathway waveforms. Visual observations on soybean pods indicated that bugs were forcing the head downward and the stylets deep into plant tissue, then pulling them up (although still remaining within the tissue), and repeating this behavior multiple times. Based on this behavior and histological images, Lucini et al. (2016) suggest that Pg1d represents laceration of a rigid cell layer (sclerenchyma cells) during pathway phase, before bugs can reach the seeds.

On soybean plants, *P. guildinii* ingest cell contents from two distinct plant tissues, xylem and seed endosperm. The xylem waveform (Pg2) is histologically correlated with stylet tips terminating in xylem vessels. Pg2 was recorded in all plant structures evaluated (leaflet, stem, and pod) and presented similar characteristics and appearance between them. Pg2 is characterized by a highly regular waveform composed of repetitive wave portions interspersed with regularly distributed peaks (Lucini et al. 2016) (see Fig. 4.4, Sect. 4.3).

A short interruption waveform (interruption phase) was observed during xylem ingestion (Pg4), mostly near the beginning of this waveform. Pg4 is a plateau waveform with peaks and an irregular appearance (Lucini et al. 2016). It probably represents watery saliva secretion for tasting/testing xylem (Backus et al. 2013).

Pg3 is only observed on soybean pods and represents feeding activities in seeds. It is divided into sub-types Pg3a and Pg3b, which occur interspersed over the entire Pg3 waveform. Pg3a presents an irregular appearance with peaks irregularly spread; Pg3b is highly regular and of short duration (a few seconds) compared with Pg3a (Lucini et al. 2016). Visual observations during Pg3a show the stink bugs moving their stylets in and out quickly and continuously in the pod tissue, which represents cell rupture feeding activities.

After a Pg3a event, the stylets remain briefly motionless inside pod tissue, resulting in waveform Pg3b, representing short ingestion events from seed endosperm previously degraded by cell rupture activities. These activities have the potential to cause severe damage in the plant tissue attacked; fresh cuts of soybean pods after a

Pg3 event show an extensive damaged area (white region) in seed endosperm (Lucini et al. 2016).

4.2.8 Tibraca limbativentris

The rice stalk stink bug *T. limbativentris* is a pentatomid occurring widely in South American countries, where it is considered a key pest on the rice crop (Pantoja et al. 1995; Fernandes 1998). *T. limbativentris* is a potential invasive species in the USA (Brambila 2007). This species feeds exclusively on vegetative structures (stems) of rice in any developmental plant stage. During early plant development (seedlings), feeding in the stems results in the symptom known as "dead-heart" (central leaf dies); during the reproductive period, damage causes the so-called white-panicle, which results in malformed panicles (Krinski and Foerster 2017).

Feeding behavior of *T. limbativentris* adult females was recorded on stalks of rice plants during vegetative plant development (V5 stage – collar formation on fifth leaf on main stem [Counce et al. 2000]) using a four-channel AC-DC monitor. Six different waveform types/subtypes were generated; two represent non-feeding activities (Z and Np) and four represent feeding activities (Tl1, Tl2, Tl3a, and Tl3b) (Almeida et al. 2020). These waveforms and their proposed biological meanings are summarized in Table 4.8.

Two non-feeding waveforms (Z and Np) were recorded and visually correlated with their biological activities. Z displays as a flat waveform with slight or no variation in appearance and represents the baseline of the recording. This wave represents the bug resting on the plant surface. In contrast, Np presents high and irregular spikes, which is correlated with the bug moving on the plant surface (Almeida et al. 2020).

Waveform type Tl1 characterizes the pathway phase. This waveform presents an irregular appearance with several peaks irregularly spread (Almeida et al. 2020). Histological observations indicate that Tl1 represents the initial stylet penetration, secretion of a salivary sheath (complete or incomplete depending on feeding strat-

Table 4.8 Summary of the EPG waveforms of *Tibraca limbativentris* adults recorded on rice stalk (V5 stage) and their proposed biological activities

Phase	Family	Type or subtype	Proposed biological activities
Non-feeding	–	Z	No movement on the stalk surface
		Np	Walking on the stalk surface
Pathway	P	Tl1	Stylet penetration and salivary sheath secretion
Ingestion	I	Tl2	Xylem sap ingestion
Salivation/ingestion	I	Tl3a	Cell laceration, enzymatic maceration of stalk tissues, and ingestion
Ingestion	I	Tl3b	Short ingestion of macerated tissues

Source: adapted from Almeida et al. (2020)

egy used), and deeper stylet penetration into plant tissue towards the feeding site. Almeida et al. (2020) suggest that *T. limbativentris* use both feeding strategies while feeding on rice stalks, the salivary sheath while ingesting from xylem vessels and cell rupture while ingesting parenchyma tissue. A similar behavior is also reported for *D. melacanthus*, which also uses both feeding strategies while feeding on maize seedlings (Lucini and Panizzi 2017a).

The ingestion phase comprises two distinct waveforms, Tl2 and Tl3. Tl2 is highly regular with rounded waves interspersed by peaks regularly distributed over the recording. Histological sections indicate that Tl2 represents xylem sap ingestion (Almeida et al. 2020). Tl3 is divided into subtypes Tl3a and Tl3b, which occur interspersed. Tl3a is primarily characterized as highly irregular containing peaks randomly distributed; Tl3b presents a regular pattern with shorter duration compared to Tl3a (Almeida et al. 2020).

During Tl3a bugs move their stylets in and out quickly inside the rice stalk. This well-known behavior represents the laceration tactic as part of the cell rupture strategy (Backus et al. 2005b). Almeida et al. (2020) proposed that Tl3a represents the mechanical action of the stylets (laceration) to disrupt the parenchyma cells. However, they suggest that the bug probably uses maceration as well, to chemically degrade the cells for subsequent ingestion. For Tl3b the stylets remain briefly (few seconds) motionless inside the plant tissue. Almeida et al. (2020) propose that Tl3b represents the ingestion of the cell contents previously degraded via laceration/maceration activities. Cross sections of fresh rice stalks after a Tl3 event show an area of damage (an opaque white region) attributed to the cell rupturing activities. Indeed, the symptoms observed in the field, known as "dead-heart" and "white-panicle," may be the result of such damage, following a long period of cell rupture feeding (Almeida et al. 2020).

4.3 Comparison of EPG Waveforms Within Pentatomidae

The EPG waveforms recorded for each feeding activity/tissue, in general, are similar between pentatomid species studied. In summary, those waveforms can be grouped into four main phases: non-feeding, pathway, X wave, and salivation/ingestion. In turn, the salivation/ingestion phase can be grouped into four waveform types based on the most common ingestion tissue: xylem, phloem, parenchyma tissue, and seed endosperm. We will compare the waveforms related to each type of phase and ingestion site among all stink bug species evaluated via EPG, highlighting their similarities and differences.

Non-feeding Phase In all pentatomid species evaluated via EPG, the non-feeding behavior represents the most dominant activities (longest total duration) among all waveforms recorded. For example, *D. furcatus* spent ca. 60–85% of their time on non-feeding activities when provided with soybean pods (preferred food source) and seed heads of spring cereals (Lucini and Panizzi 2020). The non-feeding waveform types are similar in appearance among species and share the same biological meanings. The most frequent activities reported are related to insect resting (standing still) and moving (walking, grooming) on the surface of their hosts.

Other non-feeding activities are also reported; for example, the labial dabbing activity reported for *N. viridula* (Mitchell et al. 2018) and *P. guildinii* (Lucini et al. 2016) on soybean plants. These waveforms are analogous and represent the bug tapping the plant surface with the labium, with or without producing watery saliva, before insertion of stylets into the plant. Epidermal scanning (different from labial dabbing) is another non-feeding behavior observed on *H. halys*, where the labium is dragged on plant surface to maybe break the epidermal cells and facilitate selection of a penetration site (Serteyn et al. 2020). Oviposition has also been reported via EPG, which was observed on *D. melacanthus* on maize seedlings (Lucini and Panizzi 2017a) and *E. heros* on soybean pods (TL, personal communication), as well as the excretion activities of *E. heros* observed on soybean (Lucini and Panizzi 2018b) (see Chap. 8).

Pathway Phase All pentatomids perform pathway waveforms, whether feeding on vegetative or reproductive structures of their hosts. In four stink bug species (*D. melacanthus*, *E. meditabunda*, *H. halys*, and *T. limbativentris*), only one pathway waveform was identified; in two species (*D. furcatus* and *N. viridula*), two distinct pathway waveforms (subtypes) were separated, and in the remaining two species (*E. heros* and *P. guildinii*), four different subtypes were distinguished. Independent of the number of pathway subtypes identified, pathway patterns are similar among pentatomids and primarily consist of highly variable and irregular waveforms containing randomly distributed peaks. Pathway waveforms recorded from *E. heros* are presented in Fig. 4.1 for illustration.

The pathway phase represents multiple activities performed by stylets outside of the actual ingestion site. These activities include initial and deep stylet penetration into the plant tissue and secretion of gelling saliva to create a branched or unbranched salivary sheath, which may be observed in histological sections (Fig. 4.2); for further discussion and illustration of histology, see Sect. 6.7 in Chap. 6.

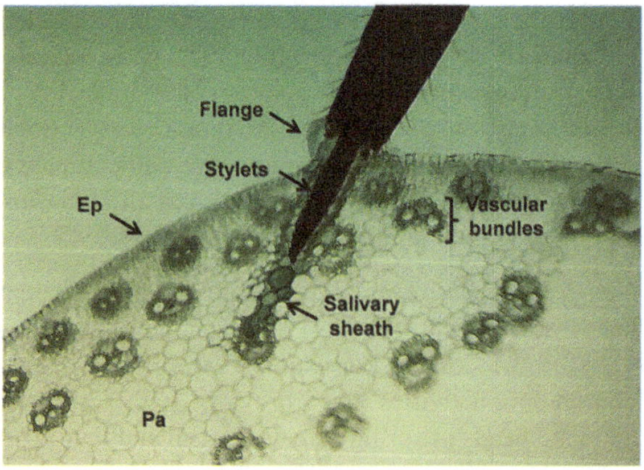

Fig. 4.2 Histological section of a maize seedling containing the severed stylets of the stink bug *Diceraeus* (*Dichelops*) *melacanthus* (Dallas). Detail of the salivary sheath secreted around the stylets during the pathway activity until they reach the vascular bundle. *Ep* epidermis of the stem, *Pa* parenchyma. (Source: Lucini and Panizzi (2018a))

For *D. furcatus*, *E. heros*, and *P. guildinii*, one of their pathway waveform subtypes (Df1b, Eh1c, and Pg1d, respectively) consists of a stereotypical pattern with a more regular appearance compared with the other pathway subtypes (Fig. 4.1c). These three waveforms are similar, and they are observed when bugs feed on reproductive structures of host plants (wheat seed head and soybean pods), mainly preceding the feeding activities in seeds. As previously discussed, these waveforms are visually correlated with head movements that force the stylets down and then up inside the plant tissue. However, this head movement does not resemble the head and stylet movements performed by the bug while using the laceration activities of the cell rupturing. Thus, this behavior indicates the presence of a rigid cell layer making stylet penetration difficult (Lucini and Panizzi 2018a, b).

X Wave Phase X waves are species-specific waveforms that represent multiple behaviors occurring as the insect reaches the preferred ingestion tissue. This transition waveform has been reported for *E. meditabunda* on soybean stems and *H. halys* on faba bean leaves. The *E. meditabunda* X wave is recorded only preceding sustained phloem sap ingestion, whereas *H. halys* performs two distinct X waves, one preceding ingestion from phloem and the other from xylem. The *E. meditabunda* X wave is marked by a highly regular pattern (Fig. 4.3b), whereas, the putative *H. halys* phloem X wave displays an irregular appearance, overall resembling an S-shaped version of the pathway waveform Hh2.

No other pentatomids have been recorded ingesting from phloem. Nevertheless, among the many stink bug species that hydrate from xylem, no X waves have been reported preceding the ingestion waveform except in *H. halys*. Serteyn et al. (2020) observed that sustained ingestion from xylem and phloem for *H. halys* is not always preceded by an X wave. Thus, this transitional waveform might be facultative for true bugs and not always requiring testing/tasting of the cells for final acceptance, different than reported for sharpshooters (Backus et al. 2009). The *E. meditabunda* X wave is always recorded preceding sustained phloem ingestion, but this species is

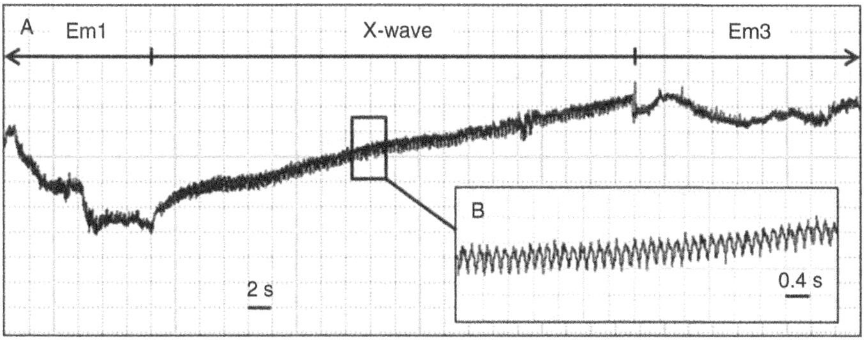

Fig. 4.3 Waveforms recorded from *Edessa meditabunda* on soybean stem (DC applied signal and input impedance of 10^9 Ohms). Overview of a feeding section showing the pathway waveform (Em1), transition waveform (X wave), and the phloem ingestion waveform (Em3) (**a**). Expanded view of the X wave (**b**)

exclusively a stem feeder (Lucini and Panizzi 2016), unlike *H. halys*. Perhaps, other pentatomids produce xylem X waves during their recordings, but maybe few individuals showed it and the researcher might have ignored it. Moreover, the X wave could be similar to xylem ingestion and/or even interruption waveforms, making its detection to pass unnoticed during analysis.

Salivation/Ingestion Phase (Vascular Tissues) Phytophagous pentatomids have, in general, a strong preference for feeding on seeds, which are rich in essential nutrients (Slansky and Panizzi 1987). However, some species feed exclusively on vascular tissues (e.g., *E. meditabunda*). All stink bug species so far evaluated via EPG ingested sap from xylem vessels on vegetative and/or reproductive structures of their hosts. Sap from vascular tissues, mostly xylem, is rich in water and poor in nutrients (Taiz and Zeiger 2004). Thus, pentatomids probably ingest sap from xylem for hydration, as reported for aphids (Spiller et al. 1990) and stink bug first instars (Rivera and Mitchell 2020) and/or for nutrient balance when ingesting from a highly concentrated reserve tissue – seeds (Lucini et al. 2016). In addition, xylem ingestion is observed preceding seed feeding in most pentatomid species; maybe this additional hydration is required before feeding in a highly concentrated tissue (Lucini and Panizzi 2018a) or might be required after being subjected to a long starvation period before recordings (see Chap. 6 for further discussion on starvation effects).

Considering all feeding waveforms recorded for pentatomids, the xylem waveforms are the most consistently similar in appearance and electrical characteristics between species. Xylem ingestion is represented by a highly stereotypical waveform composed of repetitive plateaus (arranged in rounded waves) interspersed with peaks (Fig. 4.4). These peaks are, in general, regularly distributed over time and are negatively or positively oriented varying according to stink bug species and input impedance applied during recordings.

Phloem activities are reported in only two pentatomid species (*E. meditabunda* and *H. halys*). Although both stink bugs were evaluated using the same model of DC monitor, the phloem ingestion waveforms do not share similarities. The phloem waveform from *E. meditabunda* shows a very stereotypical pattern composed of a sinuous wave, alternating large plateaus up and down, with each plateau containing an underlying, highly regular waveform, with low amplitude (Fig. 4.5), and without the typical peaks observed during xylem ingestion. This sinuous pattern resembles the sinuous changes in voltage levels during phloem ingestion from aphids, which is caused by the cell membrane depolarization of the sieve element (Salvador-Recatalà et al. 2014).

In contrast, the putative phloem waveform of *H. halys* does not show the undulation form but presents a regular waveform with peaks regularly distributed. For *E. meditabunda*, the correlation between phloem ingestion and the respective waveform was primarily done via histological observations but also indirectly via observation of droplets excreted from the severed stylets. In contrast, for *H. halys* neither histological sections nor another method was performed by Serteyn et al. (2020) to correlate the waveform with phloem ingestion. In fact, the waveform hypothesized

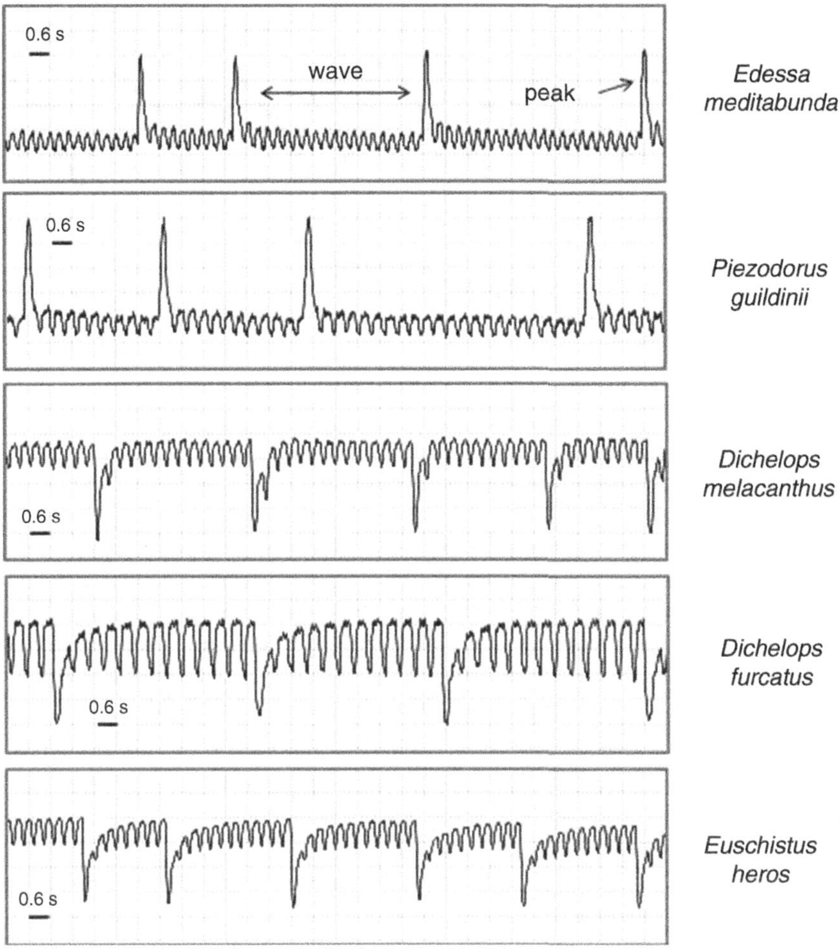

Fig. 4.4 Overview of xylem waveforms recorded from selected pentatomid species on different host plants

to represent phloem activities is similar to the xylem waveform; they differ in the peak orientation and relative position above the baseline.

During ingestion on vascular tissues, short interruptions can be observed in some stink bug species, but not all insects recorded from a species perform them. Interruptions might represent a quick watery salivation into the cells to taste/test them, as reported during xylem ingestion in sharpshooters (Backus et al. 2005a) and blissids (Backus et al. 2013). For *E. meditabunda, E. heros*, and *P. guildinii*, interruptions were recorded during xylem ingestion waveforms; *E. meditabunda* also produces these waveforms during phloem ingestion. In general, the interruption waveforms are composed of a plateau with an irregular frequency and spikes. Xylem interruptions are similar among stink bug species and are primarily recorded near

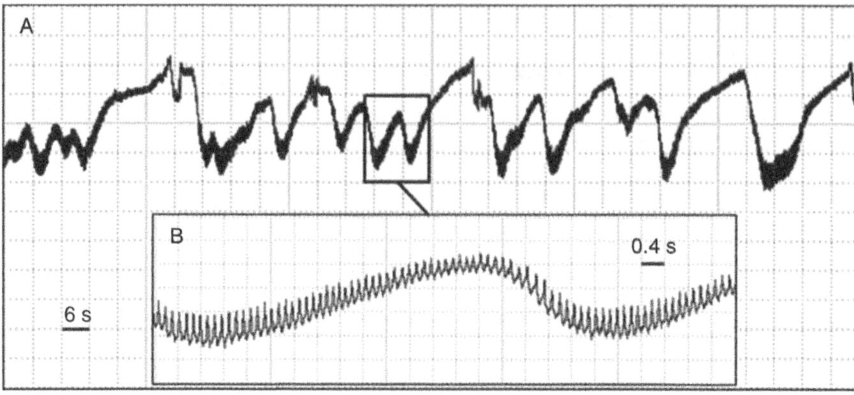

Fig. 4.5 Overview of a feeding section recorded from *Edessa meditabunda* on soybean stem (DC applied signal and input impedance of 10^9 Ohms). A phloem ingestion waveform (Em3) showing the sinuous form (**a**). Expanded view displaying the fine structure of waveform Em3 (**b**)

the beginning of xylem ingestion. In contrast, phloem interruptions from *E. meditabunda* are observed throughout the entire ingestion period (Lucini and Panizzi 2016, 2018b; Lucini et al. 2016). Perhaps, all stink bugs perform interruption waveforms during ingestion from vascular tissues, but the lack of this waveform in some stink bug species might be related to the very short duration of this waveform which could be easily overlooked or pass unnoticed during waveform analysis.

Salivation/Ingestion Phase (Parenchyma Tissue and Seed Endosperm) All pentatomids evaluated, except *E. meditabunda*, feed on parenchyma/mesophyll tissue and/or seed reserve tissue during feeding activities on their host plants. To feed on these tissues, stink bugs use the laceration and maceration tactics of the cell rupture feeding strategy to degrade a pocket of cells and then ingest the "slurry" of cell contents (Lucini and Panizzi 2018a; Mitchell et al. 2018; Almeida et al. 2020). Pentatomids are reported to use both tactics simultaneously to feed on vegetative and/or reproductive structures of their host plants (see Lucini and Panizzi 2018a and Chap. 2 for further discussion).

Interestingly, the cell rupturing waveforms recorded during feeding activities of stink bugs on vegetative (stem) and reproductive structures (seed) are similar in appearance (compare Fig. 4.6a and b) and electrical characteristics. Furthermore, waveform analysis of the cell rupture strategy has separated the laceration/maceration tactics from actual ingestion, which occur primarily interspersed with each other during the recordings (Fig. 4.6). The laceration/maceration activities are characterized by an irregular pattern containing multiple positive- and/or negative-going peaks with varying amplitude. In contrast, the ingestion waveform has a highly regular pattern (sometimes, peaks are observed) with a very short duration compared to the time spent in laceration/maceration (Lucini and Panizzi 2018a and references therein). For example, of the total time spent by *D. furcatus* on cell rupture activities on soybean seeds and spring cereals, over 70% represents lacerating/mac-

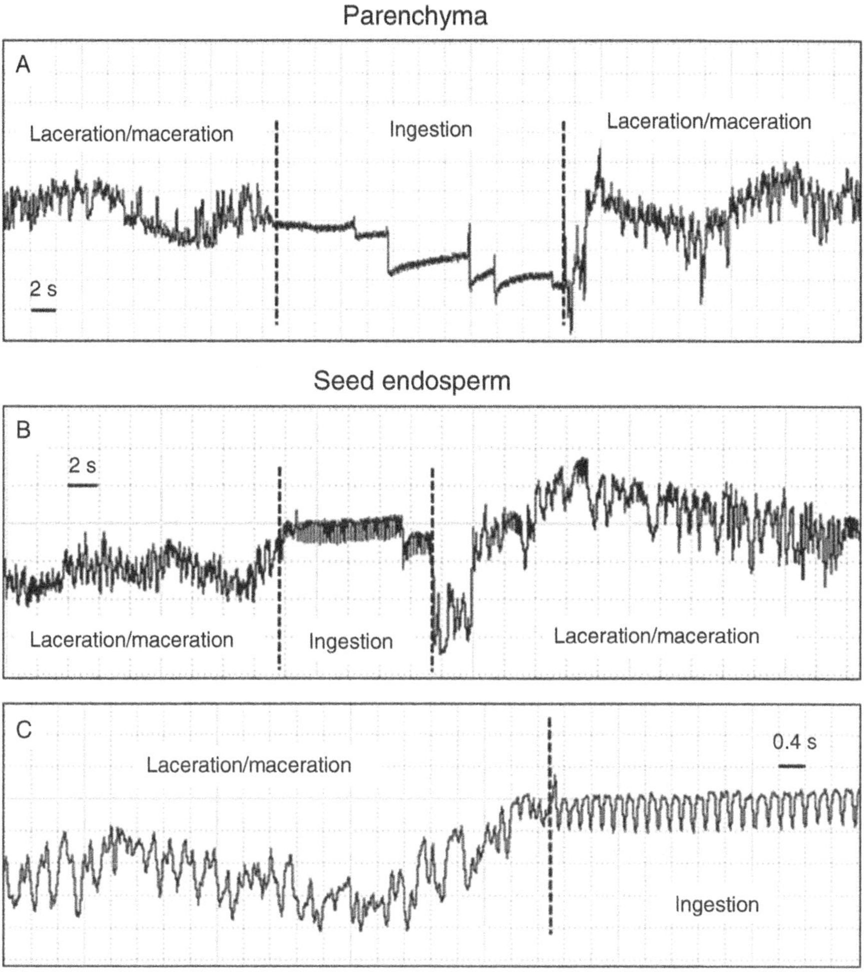

Fig. 4.6 Overview of the waveforms related to cell rupture feeding of *Diceraeus* (*Dichelops*) *furcatus* on stem of wheat seedlings (**a**) and *Euschistus heros* on soybean pods (**b** and **c**). Detail of the two waveform subtypes (cell laceration/maceration and ingestion) that occur interspersed with each other. Black dashed lines separate each stylet activity

erating activities and less than 30% ingestion, reaching only 6% on soybean seeds (Lucini and Panizzi 2020).

The ingestion-related waveform on parenchyma tissue and seed endosperm, in some cases, cannot be clearly distinguished and recorded, even though a laceration/maceration process is recorded (Lucini and Panizzi 2018a). For example, *N. viridula* on soybean seeds (Mitchell et al. 2018) and *H. halys* on faba bean leaves (Serteyn et al. 2020) do not appear to present a visible and regular waveform representing the ingestion of cell contents per se. Therefore, ingestion may be occurring simultaneously with laceration/maceration activities during very short events not distinguishable to the eye (Lucini and Panizzi 2018a).

4.4 Concluding Remarks

We have compiled and briefly discussed the majority of the studies so far performed with phytophagous stink bugs (Pentatomidae) using the EPG. As pentatomids continue to gain a growing importance as pests worldwide (e.g., Schaefer and Panizzi 2000; McPherson and McPherson 2000; McPherson 2018), mitigation of their impact to crops is becoming essential. To reach that goal, and to devise sustainable integrated pest management programs, we cannot continue to rely on hard chemicals for control. In order to develop more ecologically compatible control strategies, we need to investigate and better understand the feeding process and the resulting plant damage. Studies using EPG certainly will help to achieve these goals. The promising role of EPG in developing and assessing new control methods for pentatomids and other heteropteran crop pests is further explored in Chap. 7.

References

Almeida ACS, Jesus FG, Barrigossi JAF (2020) Unveiling the feeding behavior of *Tibraca limbativentris* (Hemiptera: Pentatomidae) on rice using an electropenetrography waveform library. J Insect Sci 20:14;1–8

Backus EA (1988) Sensory systems and behaviours which mediate hemipteran plant feeding: a taxonomic overview. J Insect Physiol 34:151–165

Backus EA, Bennett WH (2009) The AC-DC correlation monitor: new EPG design with flexible input resistors to detect both R and emf components for any piercing-sucking hemipteran. J Insect Physiol 55:869–884

Backus EA, Habibi J, Yan F, Ellersieck M (2005a) Stylet penetration by adult *Homalodisca coagulata* on grape: electrical penetration graph waveform characterization, tissue correlation, and possible implications for transmission of *Xylella fastidiosa*. An Entomol Soc Am 98:787–813

Backus EA, Serrano MS, Ranger CM (2005b) Mechanisms of hopperburn: an overview of insect taxonomy, behavior, and physiology. Annu Rev Entomol 50:125–151

Backus EA, Holmes WJ, Schreiber F, Reardon BJ, Walker GP (2009) Sharpshooter X wave: correlation of an electrical penetration graph waveform with xylem penetration supports a hypothesized mechanism for *Xylella fastidiosa* inoculation. An Entomol Soc Am 102:847–867

Backus EA, Rangasamy M, Stamm M, McAuslane HJ, Cherry R (2013) Waveform library for chinch bugs (Hemiptera: Heteroptera: Blissidae): characterization of electrical penetration graph waveforms at multiple input impedances. An Entomol Soc Am 106:524–539

Backus EA, Cervantes FA, Guedes RNC, Li AY, Wayadande AC (2019) AC–DC electropenetrography for in-depth studies of feeding and oviposition behaviors. Ann Entomol Soc Am 112:236–248

Bonjour EL, Fargo WS, Webster JA, Richardson PE, Brusewitz GH (1991) Probing behavior comparisons of squash bugs (Heteroptera: Coreidae) on cucurbit hosts. Environ Entomol 20:143–149

Brambila J (2007) Heteroptera of concern to Southern US. Invasive Arthropod Workshop, Southern Plant Diagnostic Network, Clemson, South Carolina

Bundy CS, Esquivel JF, Panizzi AR, Eger JE, Davis JA, Jones WA (2018) *Piezodorus guildinii* (Westwood). In: McPherson JE (ed) Invasive stink bugs and related species (Pentatomoidea): Biology, higher systematics, semiochemistry, and management. CRC Press, Boca Raton, pp 425–451

Cervantes FA, Backus EA, Godfrey L, Akbar W, Clark TL (2016) Characterization of an EPG waveform library for pre-reproductive adult *Lygus lineolaris* and *L. hesperus* feeding on cotton squares. Ann Entomol Soc Am 109:684–697

Chuche J, Backus EA, Thiéry D, Sauvion N (2017) First finding of a dual-meaning X wave for phloem and xylem fluid ingestion: characterization of *Scaphoideus titanus* (Hemiptera: Cicadellidae) EPG waveforms. J Insect Physiol 102:50–61

Cooke SB (2014) Probing behavior of southern green stink bug, *Nezara viridula* (Hemiptera: Pentatomidae), on the soybean plant, *Glycine max*. Graduate Thesis, Winthrop University, Rock Hill, South Carolina

Counce PA, Keisung TC, Mitchell AL (2000) A uniform and adaptive system for expressing rice development. Crop Sci 40:436–443

Esquivel JF, Musolin DL, Jones WA, Rabitsch W, Greene JK, Toews MD, Schwertner CF, Grazia J, McPherson RM (2018) *Nezara viridula* (L.). In: McPherson JE (ed) Invasive stink bugs and related species (Pentatomoidea): biology, higher systematics, semiochemistry, and management. CRC Press, Boca Raton, pp 351–423

Fehr WR, Caviness CE, Burmood DT, Pennington JS (1971) Stage of development descriptions for soybeans, *Glycine max* (L.) Merrill. Crop Sci 11:929–930

Fernandes JAM (1998) Revision of the genus *Tibraca* Stål (Heteroptera: Pentatomidae: Pentatominae). Rev Bras Zool 15:1049–1060

Hamilton GC, Ahn JJ, Bu W, Leskey TC, Nielsen AL, Park YL, Rabitsch W, Hoelmer KA (2018) *Halyomorpha halys* (Stål). In: McPherson JE (ed) Invasive stink bugs and related species (Pentatomoidea): biology, higher systematics, semiochemistry, and management. CRC Press, Boca Raton, pp 243–292

Hollay ME, Smith CM, Robinson JF (1987) Structure and formation of feeding sheaths of rice stink bug (Heteroptera: Pentatomidae) on rice grains and their association with fungi. Ann Entomol Soc Am 80:212–216

Inkley DB (2012) Characteristics of home invasion by the brown marmorated stink bug (Hemiptera: Pentatomidae). J Entomol Sci 47:125–130

Krinski D, Foerster LA (2017) Damage by *Tibraca limbativentris* Stål (Pentatomidae) to upland rice cultivated in Amazon rainforest region (Brazil) at different growth stages. Neotrop Entomol 46:107–114

Large EC (1954) Growth stages in cereals. Illustration of the Feekes scale. Plant Pathol 3:128–129

Lucini T, Panizzi AR (2016) Waveform characterization of the soybean stem feeder *Edessa meditabunda* (F.) (Hemiptera: Heteroptera: Pentatomidae): overcoming the challenge of wiring pentatomids for EPG. Entomol Exp Appl 158:118–132

Lucini T, Panizzi AR (2017a) Feeding behavior of the stink bug *Dichelops melacanthus* Dallas on corn seedlings: an EPG analysis at multiple input impedances and histology correlation. Ann Entomol Soc Am 110:160–171

Lucini T, Panizzi AR (2017b) Probing behavior of *Dichelops furcatus* (F.) (Heteroptera: Pentatomidae) on wheat plants characterized by electropenetrography (EPG) and histological studies. J Insect Sci 17:65;1–15

Lucini T, Panizzi AR (2018a) Electropenetrography (EPG): a breakthrough tool unveiling stink bug (Pentatomidae) feeding on plants. Neotrop Entomol 47:6–18

Lucini T, Panizzi AR (2018b) EPG monitoring of the Neotropical brown-stink bug, *Euschistus heros* (F.), on soybean pods: an electrical penetration graph-histology analysis. J Insect Sci 18:5;1–14

Lucini T, Panizzi AR (2020) Electropenetrographic comparison of feeding behavior of *Dichelops furcatus* (Hemiptera: Heteroptera: Pentatomidae) on soybean and spring cereals. J Econ Entomol 113:1796–1803

Lucini T, Panizzi AR, Backus EA (2016) Characterization of an EPG waveform library for red-banded stink bug, *Piezodorus guildinii* (Hemiptera: Pentatomidae), on soybean plants. Ann Entomol Soc Am 109:198–210

Maskey K (2010) Comparison of electrical penetration graph waveforms of squash bug feeding on watermelon and its relatives. Master's Thesis, Oklahoma State University, Stillwater, Oklahoma

McPherson JE (2018) Invasive stink bugs and related species (Pentatomoidea): Biology, higher systematics, semiochemistry, and management. CRC Press, Boca Raton

McPherson JE, McPherson RM (2000) Stink bugs of economic importance in America north of Mexico. CRC Press, Boca Raton

Mitchell PL, Cooke SB, Smaniotto LF (2018) Probing behavior of *Nezara viridula* on soybean: Characterization and comparison of electrical penetration graph (EPG) waveforms on vegetative and reproductive plant structures. J Agric Urban Entomol 34:19–43

Nielsen AL, Hamilton GC (2009) Life history of the invasive species *Halyomopha halys* (Hemiptera: Pentatomidae) in Northeastern United States. Ann Entomol Soc Am 102:608–616

Panizzi AR (1997) Wild hosts of pentatomids: ecological significance and role in their pest status on crops. Annu Rev Entomol 42:99–122

Panizzi AR (2015) Growing problems with stink bugs (Hemiptera: Heteroptera: Pentatomidae): species invasive to the U.S. and potential neotropical invaders. Am Entomol 61:223–233

Panizzi AR, Lucini T (2016) What happened to *Nezara viridula* (L.) in the Americas? Possible reasons to explain populations decline. Neotrop Entomol 45:619–628

Panizzi AR, Slansky F (1985) Review of phytophagous pentatomids (Hemiptera: Pentatomidae) associated with soybean in the Americas. Fla Entomol 68:184–214

Panizzi AR, Corrêa BS, Gazzoni DL, Oliveira EB, Newman GG, Turnipseed SG (1977) Insects of soybean in Brazil. Embrapa Soja, Londrina, Bol Tec 1:1–20 (*in Portuguese*)

Panizzi AR, Bueno AF, Silva FAC (2012) Insects that attack pods and grains. In: Hoffmann-Campo CB, Corrêa-Ferreira BS, Moscardi F (eds) Soybean: Integrated management of insects and other pest arthropods. Embrapa Editora, Brasília, pp 335–398 (*in Portuguese*)

Panizzi AR, Agostinetto A, Lucini T, Pereira PRVS (2016) Effect of green-belly stink bug, *Dichelops furcatus* (F.) on wheat yield and development. Crop Prot 79:20–25

Panizzi AR, Lucini T, Possebom T (2018) Development of *Dichelops furcatus* (Hemiptera: Heteroptera: Pentatomidae) reared on spring cereals versus soybean. J Insect Sci 18:17;1–7

Pantoja A, Daza E, Garcia C, Mejia OI, Rider DA (1995) Relative abundance of stink bugs (Hemiptera: Pentatomidae) in Southwestern Colombia rice fields. J Entomol Sci 30:463–467

Prado E, Tjallingii WF (1994) Aphid activities during sieve element punctures. Entomol Exp Appl 72:157–165

Ritchie S, Hanway JJ (1989) How a corn plant develops. Iowa, Iowa State University of Science and Technology, Cooperative Extension Service, Ames, Special Report 48:1–24

Rivera KL, Mitchell PL (2020) Probing behavior of *Nezara viridula* first instars: EPG analysis and effect of food availability on subsequent development. J Agric Urban Entomol 36:47–63

Salvador-Recatalà V, Tjallingii WF, Farmer EE (2014) Real time, in vivo intracellular recordings of caterpillar-induced depolarization waves in sieve elements using aphid electrodes. New Phytol 203:674–684

Schaefer CW, Panizzi AR (2000) Heteroptera of economic importance. CRC Press, Boca Raton

Serteyn L, Ponnet L, Backus EA, Francis F (2020) Characterization of electropenetrography waveforms for the invasive heteropteran pest, *Halyomorpha halys*, on *Vicia faba* leaves. Arthropod Plant Interact 14:113–126

Silva FAC, Silva JJ, Depieri RA, Panizzi AR (2012) Feeding activity, salivary amylase activity, and superficial damage to soybean seed by adult *Edessa meditabunda* (F.) and *Euschistus heros* (F.) (Hemiptera: Pentatomidae). Neotrop Entomol 41:386–390

Slansky F Jr, Panizzi AR (1987) Nutritional ecology of seed-sucking insects. In: Slansky F Jr, Rodriguez JG (eds) Nutritional ecology of insects, mites, spiders, and related invertebrates. Wiley, New York, pp 283–320

Smaniotto LF, Panizzi AR (2015) Interactions of selected species of stink bugs (Hemiptera: Heteroptera: Pentatomidae) from leguminous crops with plants in the neotropics. Fla Entomol 98:7–17

Spiller NJ, Koenders L, Tjallingii WF (1990) Xylem ingestion by aphids – a strategy for maintaining water balance. Entomol Exp Appl 55:101–104

Taiz L, Zeiger E (2004) Plant Physiology, 3rd edn. Sinauer Associates Inc., Publishers, Sunderland

Tjallingii WF (1988) Electrical recording of stylet penetration activities. In: Minks AK, Harrewjin P (eds) Aphids: their biology, natural enemies and control. Elsevier, Amsterdam, pp 95–108

Todd JW (1989) Ecology and behavior of *Nezara viridula*. Annu Rev Ent 34:273–292

Wiman NG, Walton VM, Shearer PW, Rondon SI (2014) Electronically monitored labial dabbing and stylet 'probing' behaviors of brown marmorated stink bug, *Halyomorpha halys*, in simulated environments. PloS One 9:e113514;1–24

Chapter 5
Electronic Monitoring of the Feeding Behavior of Blissidae, Coreidae, Miridae, and Plataspidae

Contents

Abstract In this chapter, we summarize and discuss the most relevant feeding activities and their respective EPG waveforms performed by true bugs from four families, Miridae (infraorder Cimicomorpha), Blissidae, Coreidae, and Plataspidae (infraorder Pentatomomorpha). Recordings performed with species of Blissidae, Coreidae, and Plataspidae share similar waveforms and strong similarities of feeding activities, such as feeding on vascular tissues (mostly phloem cells) and use of the salivary sheath feeding strategy. In contrast, species of Miridae evaluated with EPG do not feed on vascular tissues, and none secrete a salivary sheath. To feed, they use a different feeding strategy – cell rupture – to degrade plant tissue for subsequent ingestion of cell contents. Pentatomidae, covered in Chap. 4 and herein compared with the other families, employ both feeding strategies to obtain nutrients from host plants. They share similarly shaped waveforms and biological activities with species of the four other families that have been evaluated using EPG.

Keywords True bugs · EPG · Feeding behavior · Waveforms · Feeding sites

A. R. Panizzi et al., *Electronic Monitoring of Feeding Behavior of Phytophagous True Bugs (Heteroptera)*, Entomology in Focus 6,
https://doi.org/10.1007/978-3-030-64674-5_5

5.1 Introduction

To date, 18 true bug species from 5 families have been recorded using electronic monitoring. In this chapter, we summarize the most relevant information related to waveforms and behaviors of species belonging to families Miridae (infraorder Cimicomorpha), Blissidae, Coreidae, and Plataspidae (infraorder Pentatomomorpha). During feeding activities, phytophagous mirids produce watery saliva, which contains digestive enzymes secreted to degrade plant tissue, but do not secrete gelling saliva to create a salivary sheath (Miles 1972). In contrast, phytophagous pentatomomorphans do secrete gelling saliva to form a salivary sheath as they insert the stylets; they also produce watery saliva although the enzyme content differs from that of mirids (Hori 2000). See Chap. 2 for further discussion of heteropteran saliva.

Miles (1972) associated the majority of pentatomomorphans with the salivary sheath strategy (feeding on vascular tissues), except seed feeders which were associated with lacerate-and-flush feeding; nowadays we know that pentatomid seed feeders use maceration activities as well. Cimicomorphans (Miridae) were originally associated with lacerate-and-flush feeding (mechanical action); later, Miles and Taylor (1994) reclassified them as macerate-and-flush feeders (enzymatic activities). Members of Blissidae and Plataspidae seem to be exclusively salivary sheath feeders, exploiting both types of vascular tissues, but preferentially phloem cells. Coreids from four tribes are associated with osmotic pump feeding, creating osmotic gradients to induce phloem unloading for subsequent ingestion (Miles and Taylor 1994). See Chap. 2 for further explanation of heteropteran feeding strategies.

Feeding behavior of heteropterans is complex and highly variable at all hierarchical levels of classification (infraorders, families, even species). Electronic monitoring is a potential tool to broaden knowledge, allowing exploration of behaviors not yet completely understood. In this chapter, we will compile and briefly discuss the feeding behavior of true bug species evaluated via electrical penetration graph (EPG), from four families, Blissidae, Coreidae, Miridae, and Plataspidae, summarizing the waveforms and their correlated feeding activities. Even though a waveform library at different Ri levels was created in some studies, herein, the summary of the waveform appearance and details will focus on Ri of 10^7 Ohms, as that is, in general, the optimal Ri level for heteropterans. Finally, we will compare EPG studies across all families, including those conducted with the most studied family – Pentatomidae – covered in Chap. 4.

5.2 EPG Research on Blissidae

True bugs of the genus *Blissus*, commonly known as chinch bugs, are phytophagous, feeding on several species of graminaceous plants (Poaceae), mostly cultivated grasses. They have been considered important economic pests in the USA for a long time (Reinert and Kerr 1973). Among *Blissus* species, two congeneric species have gained importance as pests on cultivated grasses in North America, the southern chinch bug, *Blissus insularis* Barber, and the western chinch bug, *Blissus occiduus* Barber (Reinert and Kerr 1973; Baxendale et al. 1999). The latter species also feeds on barley, wheat, and sorghum (Eickhoff et al. 2004). As they feed, damaged tissue suffers discoloration, leading to yellowish/brownish patches; depending on infestation level, this may kill large areas of turf (Reinert and Kerr 1973; Baxendale et al. 1999).

Feeding behavior of adults of two chinch bug species (*B. insularis* and *B. occiduus*) was recorded on stolons of grass plants using a four-channel EPG AC-DC monitor (Backus et al. 2019; EPG Technologies, Inc., Gainesville, FL, USA). Waveforms produced by both species were similar, and nine feeding waveform types were identified: G1, G2, H, J, H-I1, H-I2, J-I1, J-I2, and N (Backus et al. 2013). These waveforms and their proposed biological meanings are summarized in Table 5.1.

Pathway phase comprises three waveforms (G1, G2, and H). Family G is divided into types G1 and G2. G1 presents a jagged appearance with several spikes/peaks; G2 is a stereotypical waveform composed of plateaus and valleys with high amplitude, but lower than G1. On top of these plateaus, multiples spikes occur; thus, they are named spiky plateaus (Backus et al. 2013).

Waveform H is preceded by either G1or G2 waves. The relative amplitude is much lower compared to G waveforms. H presents medium-structure with irregular frequency and variable shape; in addition, this waveform shows an underlying fine-structure composed of wave and spike portions. In general, the H wave represents the largest portion of pathway phase in each feeding event of blissids (Backus et al. 2013).

Table 5.1 Summary of the EPG waveforms of chinch bug adults recorded on stolons of grass plants and their proposed biological activities

Phase	Family	Type	Proposed biological activities
Pathway	G	G1	Secretion of thickest gelling saliva, formation of salivary flange and trunk
		G2	Secretion of thickest gelling saliva, formation of salivary trunk
	H		Secretion of thinner gelling saliva, deeper penetration of stylets to vascular bundle
J-wave	J		Penetration of phloem sieve element; salivation, tasting, testing of salivary seal
Ingestion	I	H-I1	Active ingestion from and possible salivation into xylem tracheary element
		H-I2	Active ingestion from xylem tracheary element
	I	J-I1	Passive/active ingestion from and possible salivation into phloem sieve element
		J-I2	Passive/active ingestion from phloem sieve element
Interruption	N		Salivation into vascular cell

Source: adapted from Backus et al. (2013)

Chinch bugs are salivary sheath feeders, and, based on recordings of other hemipterans that use this strategy, pathway waveforms undoubtedly represent gelling saliva secretion. Waveform G2 strongly resembles waveform A1 of sharpshooters, which is correlated via histology with secretion of salivary sheath flanges and trunks and deep stylet penetration (Backus et al. 2005). Waveform H also resembles another sharpshooter waveform (B1 wave), histologically correlated with secretion of thinner gelling saliva and deep stylet penetration to reach vascular vessels (Backus et al. 2005).

Chinch bugs ingest from two feeding tissues, xylem tracheary elements and phloem sieve elements, comprising four waveform types: H-I1, H-I2, J-I1, and J-I2. This behavior includes waveform I (ingestion) and J waves (proposed to represent X waves; see below). Waveform I1 is transitional and stabilizes as waveform I2. If recorded after the pathway phase H, the waveforms are termed H-I1 and H-I2 (xylem waveforms); if recorded following a J wave, the ingestion waveforms are termed J-I1 and J-I2 (phloem waveforms). In general, H-I waveforms (H-I1 and H-I2) have short event durations (<1 h) and large absolute amplitude, whereas J-I waveforms (J-I1 and J-I2) have a longer duration (>1 h) and much smaller amplitude (Backus et al. 2013).

The ingestion waveform (I) is highly regular. Usually, I1 and I2 consist of repetitive episodes (plateaus), each one starting with single peak followed by a wave portion. Type I1 is the first ingestion waveform recorded and presents a variable number and shapes of plateaus. As I1 is recorded, wave fine-structure becomes more stable in appearance for a long time (hours) (Backus et al. 2013).

Based on electrical findings and presence of X waves (described later), J-I1 represents salivation into and ingestion from phloem, whereas J-I2 represents sustained ingestion from phloem (Backus et al. 2013). J-I1 resembles the E1 waveform recorded for aphids, which is correlated with salivation into phloem tissue (Prado and Tjallingii 1994). H-I1 probably represents salivation and active ingestion from xylem tracheary elements and H-I2 represents sustained ingestion from this tissue (Backus et al. 2013). H-I2 is similar to xylem waveforms recorded in other heteropterans, composed of peaks alternating with wave portions (Lucini and Panizzi 2018a; Mitchell et al. 2018; Serteyn et al. 2020).

During xylem ingestion, an interruption waveform (named N) is observed. N is a short waveform (a few seconds), irregular, characterized by a sudden voltage increase quickly followed by abrupt voltage drop, resulting in a positive flat-spiky plateau (Backus et al. 2013). The chinch bug N waveform resembles interruption waveforms (also termed N) recorded for sharpshooters while ingesting sap from xylem vessels, which are histologically correlated with salivation into xylem cells and tasting the contents (Backus et al. 2005, 2012). Based on that similarity, Backus et al. (2013) proposed a similar behavior for the N waveform recorded from chinch bugs.

No interruption waveforms are observed during phloem ingestion (J-I wave). However, a transition waveform J is often recorded after the pathway waveform H and always preceding J-I1. The chinch bug J wave recorded at 10^9 Ohms is strongly similar to the X wave recorded from the pentatomid *Edessa meditabunda* (F.) using a DC monitor with the same input impedance; this waveform also preceded phloem ingestion activity (Lucini and Panizzi 2016).

Backus et al. (2013) proposed that the J wave from chinch bugs represents an X wave. This waveform type is species-specific and primarily reported from xylem and phloem feeders (e.g., Backus et al. 2009; Chuche et al. 2017). X waves are associated with stylet penetration into the preferred ingestion cell and subsequent activities (a combination of salivation, fluid ingestion, and egestion) to taste/test the cell contents or also to overcome vascular cell defenses (e.g., P proteins in phloem), before final acceptance or rejection of the tissue (Backus et al. 2009; Walker and Medina-Ortega 2012).

5.3 EPG Research on Coreidae

Coreids are plant feeders, with a preference for dicots (Schaefer and Mitchell 1983). Host plants include fruits, grain legumes, tree nuts, and garden vegetables; a few coreid species are reported to cause significant losses (Mitchell 2000). Coreids produce a salivary sheath and may ingest from vascular tissues (Neal 1993), hydrate from xylem (Mitchell 2006), and/or feed on fruits and seed endosperm (Mitchell 2006; Rodrigues et al. 2007). Other species feed on phloem tissues using the osmotic pump feeding strategy (Hori 2000). No EPG study has been done with a known osmotic pump feeder (see Chap. 2 for discussion of this strategy).

The first published EPG study with a heteropteran was conducted with a coreid species, with nymphs of the squash bug *Anasa tristis* (De Geer) feeding on different cucurbitaceous plants (Bonjour et al. 1991). Several years later, a second EPG paper was published, also with nymphs of the squash bug (Cook and Neal 1999). To date, *A. tristis* is the only coreid species for which EPG data have been published.

In the study conducted by Bonjour et al. (1991), feeding behavior of *A. tristis* first instars was electronically monitored on leaves of multiple cucurbits (squash, pumpkin, watermelon, cucumber, muskmelon, and buffalo gourd). In total, six feeding waveforms were identified on leaves of all hosts evaluated: one represents non-feeding (B), and five are related to feeding activities (P, E, Wa, Wb, and Wc). Histological sections were made from leaves of squash plants during waveform Wc to determine the ingestion tissue (Bonjour et al. 1991). These waveforms and their proposed biological meanings are summarized in Table 5.2.

The non-feeding phase, waveform B, is a flat line and represents the baseline of the recording. During the B waveform, insects stand still, walk on the plant surface, and display other non-feeding activities. The pathway phase is represented by two waveforms (P and E). Waveform P, the first feeding behavior, represents stylet insertion into plant tissue and always precedes waveforms Wa and Wb. It is characterized by an abrupt increase in voltage level from baseline and exhibits multiple irregular peaks. Waveform E is observed only at the end of a feeding event and represents stylet removal from plant tissue. E wave is also represented by a peak (with smaller amplitude than P wave), which decreases in voltage level as stylets are removed.

Table 5.2 Summary of the EPG waveforms of *Anasa tristis* first instars recorded on different cucurbit host plants (squash, pumpkin, watermelon, cucumber, muskmelon, and buffalo gourd) and their proposed biological activities

Phase	Waveform type	Proposed biological activities
Non-feeding	B	Insect standing still/walking
Pathway	P	Insertion of stylets into the plant tissue
	E	Stylet removal
Ingestion	Wa	Unknown
	Wb	Unknown
	Wc	Ingestion from vascular tissues

Source: based on Bonjour et al. (1991)

The ingestion phase consists of three waveforms: Wa, Wb, and Wc. Wa presents irregular peaks with high amplitude and is always recorded between two B waveform events. The remaining two ingestion waveforms, Wb and Wc, are recorded together during feeding activities with Wb always preceding Wc. Wb is composed of high peaks that decrease in amplitude and evolve into Wc. The biological meanings of Wa and Wb are unknown. For Wc, histological sections indicate ingestion from vascular tissues (either phloem or xylem vessels). In addition to these waveforms, Cook and Neal (1999) recorded labial dabbing (called "sampling"), in which a series of spikes were formed as the insect repeatedly touched its mouthparts to the plant surface.

More recently, feeding behavior of fourth and fifth instar *A. tristis* was digitally recorded on watermelon using the EPG "Missouri monitor" (Backus and Bennett 1992) linked to a computer (Maskey 2010; unpublished MS thesis), which allowed clearer and more detailed waveforms to be obtained. They were described based on *Lygus hesperus* Knight (Cline and Backus 2002) and previous *A. tristis* (Bonjour et al. 1991) waveforms as Z (non-feeding activities), A (probing, called by authors a "test probe," which we consider a redundancy), B (pathway activities with secretion of a salivary sheath), and C (ingestion from vascular tissues, probably xylem vessels, which was determined via histological observations) (Maskey 2010).

5.4 EPG Research on Miridae

Miridae is the largest and the most diverse family within Heteroptera. Phytophagy predominates, including several species reported as pests causing significant damage to crop plants around the world (Wheeler 2000a). However, many species are considered scavengers or facultative predators; some species show exclusively predatory habits, making them potential biological control agents (Wheeler 2000a,b). The majority of species are oligophagous, but some *Lygus* are extremely polyphagous, feeding on more than 100 plant species (see Wheeler 2000a and references therein).

EPG has been used to investigate feeding behavior of nymphs and adults on different food sources of six mirid species: *Apolygus lucorum* (Meyer-Dür), *L. hesperus*, *Lygus lineolaris* (Palisot de Beauvois), *Nesidiocoris tenuis* (Reuter), *Stenotus rubrovittatus* (Matsumura), and *Trigonotylus caelestialium* (Kirkaldy). Feeding behavior of *A. lucorum* was monitored on cotton leaves, on green bean pods, and on eggs of the caterpillar *Helicoverpa armigera* (Hübner). This species is considered zoophytophagous (Yuan et al. 2013). *N. tenuis* is also a zoophytophagous species (De Puysseleyr et al. 2013), but its feeding behavior was only evaluated on tomato plants. The remaining four species are phytophagous, and their feeding behavior was investigated on cotton and wheat plants. The majority of EPG studies with mirids have been performed with the tarnished plant bugs, *L. lineolaris* and *L. hesperus*, including detailed waveform descriptions as well as quantitative studies (references in Table 5.1 in Chap. 1). We therefore focus primarily on these two species.

5.4.1 Early Research with Lygus *Using the Missouri Monitor*

The first published EPG paper on mirids was performed early in the 2000s for *L. hesperus* feeding on host and non-host plants using a four-channel Missouri monitor (Cline and Backus 2002). Stylet activities and biological meanings were correlated with waveforms recorded, based on visual observations on artificial diet using videomicrography. In addition, body posture was also correlated with stylet activities using video recordings. Feeding behavior of *L. hesperus* third instars was recorded on cotton, okra, and cauliflower, and on artificial diet containing ink particles. Nine waveforms were identified, three representing non-feeding (Z1, Z2, and D) and six representing feeding activities (A, B, C1, C2, E, and F). These waveforms and their proposed biological meanings are summarized in Table 5.3.

Table 5.3 Summary of the EPG waveforms recorded from *Lygus hesperus* third instars, using a Missouri monitor, on plants (cotton, okra, and cauliflower) and artificial diet, and their proposed biological activities

Phase	Type	Proposed biological activities
Non-feeding	Z1	Any active movements performed by the insect
	Z2	Insect resting on plant surface
	D	Excretory droplets
Probing	A	Brief stylet penetration into substrate
Drilling/ laceration	B	Mechanical and chemical degradation of plant cells
Ingestion	C1	Ingestion of cell contents
	C2	Ingestion of cell contents
–	E	Deposition of saliva droplet (bubble) on substrate as stylets are removed from artificial diet
–	F	Repeated stylet insertion

Source: adapted from Cline and Backus (2002)

Non-feeding behaviors include activities without insertion of the stylets into the plant tissue. *L. hesperus* produced three non-feeding waveforms, Z1, Z2, and D. Z1 has an irregular pattern and represents any activity (movement) performed on plant or artificial diet surfaces. In total, 14 different moving activities were reported using the video correlation. Z2 displays a flat-line waveform and is correlated with the bug resting (Cline and Backus 2002). Another non-feeding behavior reported is represented by waveform D, a high amplitude and square plateau waveform. D is a very short waveform (< 3 s) and associated with excretion of liquid (anal excretory droplets) on the substrate surface (Cline and Backus 2002).

The most common feeding waveform recorded for *L. hesperus* is denoted A. This wave is irregular with brief duration (a few seconds) and composed of multiple sub-patterns. Video observations demonstrated that during waveform A, insects performed activities on plants and artificial diet related to tasting/testing food to determine its quality and palatability before final acceptance. This waveform was the only one recorded on cauliflower (non-host), which supports its biological meaning (Cline and Backus 2002).

The second most frequently recorded waveform for *L. hesperus* is B. It presents a variable appearance, composed of multiple irregular peaks. During waveform B, drilling/laceration activities occur, during which bugs degrade plant tissue mechanically and chemically for subsequent ingestion. Video recordings on artificial diet demonstrated stylets moving quickly and repeatedly within the diet, coupled with labial bending and head bobbing (Cline and Backus 2002). This behavior is similar, but not identical, to laceration waveforms recorded from *Empoasca* spp. leafhoppers (Calderon and Backus 1992; Jin et al. 2012).

Ingestion behavior comprises waveform types C1 and C2, which follow waveform B. C1 is less common and appears as a flat pattern interspersed with small peaks. C2 represents the most common ingestion waveform from *L. hesperus*. It is similar in appearance to C1 but shows higher amplitude of both wave and peak portions. Video recordings on artificial diet and plants show stylets motionless during both waveforms C1 and C2. Furthermore, on artificial diet a continuous movement of ink particles toward the stylet tips was observed, indicating ingestion behavior. Therefore, C1 and C2 recorded on plants represent ingestion of cell contents (Cline and Backus 2002). C2 is similar to waveform Ib from *Empoasca* spp. which is related to fluid uptake from mesophyll cells with quick secretion of saliva (Calderon and Backus 1992).

During feeding by *L. hesperus*, some activities were restricted to specific substrates. Waveform E, the salivary bubble, is thought to be an artifact of recording on artificial diet; it resembles waveform D but occurs as stylets are removed. Another activity is waveform F, which represents multiple, shallow stylet reinsertions. It is irregular, with high amplitude, and was recorded infrequently and only on cotton squares (buds) (Cline and Backus 2002).

5.4.2 *Recent Research with* Lygus *Using the AC-DC Monitor*

Using an AC-DC EPG monitor, Cervantes et al. (2016) obtained more detailed waveforms and developed a complete waveform library for two *Lygus* species. Feeding behavior of *L. lineolaris* and *L. hesperus* adults was recorded on pinhead cotton squares. Eight waveforms were identified, three representing non-feeding activities (S, A, and W) and five related to feeding activities (CR, T1, T2, T3, and I). These waveforms and their proposed biological meanings are summarized in Table 5.4.

For both *Lygus* species, non-feeding phase comprises three waveforms (S, A, and W), which were visually correlated with their respective biological meanings. Waveform S (termed Z2 in Cline and Backus 2002) is recorded when the bug is standing still. Waveform A, consisting of single or multiple positive high-amplitude spikes which are detected and registered at high Ri levels, correlates visually with antennation (i.e., single or repetitive touches of antennal tips to the plant surface) (Cervantes et al. 2016). Waveform W (termed Z1 in Cline and Backus 2002) represents walking on the plant surface and shows an irregular pattern with several peaks. Waveforms that represent standing still and walking have been extensively recorded via EPG, whereas antennation behavior (waveform A) was recorded for the first time (Cervantes et al. 2016) (see Chap. 8 for illustration and further discussion of waveforms of non-feeding behaviors).

Cell rupturing behavior appears as waveform CR (termed B in Cline and Backus 2002). This waveform always starts with an abrupt decrease or increase in voltage level from baseline; it has a variable appearance and amplitude and lacks discernible patterns (Cervantes et al. 2016). The CR waveform does not resemble the pathway waveform of salivary sheath feeders, which would be expected as mirids do not secrete gelling saliva (Miles 1972). Based on electrical characteristics, and loss of regularity and definition, Cervantes et al. (2016) proposed that CR represents salivation, with secretion of enzymatic watery saliva that dissolves cell contents which are eventually flushed out (maceration activity).

Table 5.4 Summary of the EPG waveforms of *Lygus lineolaris* and *L. hesperus* adults recorded, using an AC-DC monitor, on pinhead cotton squares and their proposed biological activities

Phase	Family	Type	Proposed biological activities
Non-feeding	S		No movement or activity
	A		Brief and repeated touches of antennal tips on plant surface
	W		Walking on plant surface
Cell rupturing	CR		Stylet maceration and slow stylet movement
Transition behavior	T	T1	Salivation and/or cibarial quivering
		T2	Tasting via precibarial valve fluttering
		T3	Unknown
Ingestion	I		Sustained ingestion/cibarial pumping

Source: adapted from Cervantes et al. (2016)

In some feeding sessions, the CR waveform suddenly ends and a short transition waveform (named T) is observed. This transition behavior is always recorded after CR and always precedes ingestion waveform I (described later); it is clearest when recorded at an input impedance of 10^9 Ohms. The T waveform exhibits a stereotypical pattern composed by repetitive episodes, each one containing three subtypes (T1, T2, and T3). T1 and T2 display a regular pattern although T2 has a higher frequency and lower amplitude than T1; T3 is irregular with low amplitude, a nearly flat waveform (Cervantes et al. 2016).

The repetitive and highly stereotypical episodes of T waveform resemble the X waves extensively reported for species of salivary sheath feeders on phloem and xylem (e.g., Backus et al. 2009, 2013; Lucini and Panizzi 2016; Chuche et al. 2017). However, this represents the first time an X wave was recorded for a heteropteran cell rupture feeder. This *Lygus* X wave is probably related to tasting/testing of plant tissue previously macerated to obtain the optimum level of cell liquefaction prior to ingestion (Cervantes et al. 2016).

Based on electrical characteristics and similarity to X waves described for other hemipterans, Cervantes et al. (2016) propose the following activities for each T subtype: T1 may represent rapid fluid uptake or egestion behavior plus salivation into the ingestion site to taste cell contents; T2 is thought to represent tasting of the internal fluids of the plant previously taken up during T1; and T3 may represent additional secretion of saliva.

The ingestion phase that follows is represented by waveform I (C1 and C2 in Cline and Backus 2002). For both *Lygus* species, the ingestion waveform shows a highly regular frequency composed of peak and wave portions with a stereotypical pattern. Although regular, the appearance and structure (amplitude and peak orientation) of the I waveform is affected by type of applied signal and input impedance (Cervantes et al. 2016).

The I waveform of *Lygus* spp. refers to sustained ingestion of cell contents previously degraded during cell rupturing (CR). It shares similarities (i.e., the repetitive pattern) with ingestion waveforms recorded in other heteropteran species (Backus et al. 2013; Lucini and Panizzi 2018a).

5.4.3 Recent Research with Other Mirid Species

The remaining mirid species evaluated with EPG include two zoophytophagous (*A. lucorum* and *N. tenuis*) and two phytophagous species (*S. rubrovittatus* and *T. caelestialium*). The EPG studies with these mirids are primarily comparative studies in which waveform description (electrical characteristics and details) was not emphasized. The authors relied on previous EPG studies, particularly from *Lygus* spp., to identify and quantify their waveforms. Herein, we will briefly discuss the most relevant aspects related to the waveforms and feeding activities of those mirid species.

For *A. lucorum* the EPG was used to compare the feeding behavior of nymphs on plant, insect prey, and mixture of both (plant plus insect prey) (Li et al. 2016) and recently was used to evaluate the effect of insecticide on their feeding activities (Lu et al. 2020). For both studies, *A. lucorum* was evaluated using a Giga model DC monitor (EPG Systems, Wageningen, the Netherlands) (Tjallingii 1988). Li et al. (2016) identified four main waveforms based on patterns described in *L. hesperus* in Cline and Backus (2002), which were named I, II, III, and IV; and although *A. lucorum* was evaluated on plant and insect prey (zoophytophagy), the behaviors were strongly similar with *Lygus* bugs. Waveform I was recorded after stylets had been briefly (< 3 s) inserted into the food tissue and probably represents probing. Waveform II was correlated with multiple-cell laceration, plus secretion of watery saliva to degrade the food tissue, i.e., the drilling/laceration behavior. Waveform III represents the ingestion activities of the degraded tissue. Finally, the short (< 1 s) waveform IV was recorded after wave III and represents stylet removal from tissue.

For *N. tenuis*, EPG was used to compare the feeding behavior of immature (fifth instar) and adult stages (males and females) on petioles of tomato plants at vegetative developmental stage (V5-V6) using a Giga-8 model DC monitor (Chinchilla-Ramírez et al. 2021). The authors used the waveform library developed for *Lygus* spp. (Cervantes et al. 2016) for identification and quantification of the waveforms. They evaluated two feeding waveforms, CR (cell rupture) and I (ingestion). The T transitional waveform (T) was merged with CR waveform because T was rarely recorded and not clearly distinguished. In addition, histological analysis was done to correlate waveforms CR and I with the stylet tip position in the plant tissue.

Cell rupturing activities represented the majority of the time spent by the insects, independent of the developmental stage; this behavior appeared to be the major factor causing plant damage. No salivary sheath was secreted in the plant tissue during feeding. Stylet tips were located in vascular bundles and the interfascicular parenchyma, indicating that *N. tenuis* fed in the vascular ring region (Chinchilla-Ramírez et al. 2021).

The mirids *S. rubrovittatus* and *T. caelestialium* were investigated via EPG to determine their diurnal feeding activities on wheat seedlings (Suzuki and Hori 2014) and response to rice starch and related substances using a variation of artificial diet, i.e., a moistened cotton pad containing the test substance (Hori and Naito 2018) (see Sect. 6.6 in Chap. 6 for further details on artificial diet methodology). The feeding behavior was recorded using a two-channel-type version of an EPG monitor (Tsukubarikaseiki, Tsukuba, Japan). The authors relied on the waveforms developed for *L. hesperus* (Cline and Backus 2002) for characterization of their waveforms. Suzuki and Hori (2014) classified the behaviors of these species on wheat seedlings as non-feeding activities (walking or standing still), probing, drilling, and ingestion. Subsequently Hori and Naito (2018), using artificial diet, distinguished seven waveforms and named them Np, T, D, Ig, Is, NI, and W, which included all behaviors previously described by Suzuki and Hori (2014).

The waveform Np represents the insect moving or standing still on the food substrate. Waveform T represents probing, in which the stylets penetrate the substrate for a few seconds. Waveform D is associated with drilling/laceration behavior, which includes multiple stylet activities (insertion, movements, and salivation) for degradation of the tissue. Next, the Ig waveform is observed which represents ingestion of the food material previously degraded. The remaining waveforms (Is, NI, and W) represent the "resting behavior" (not related to the resting behavior reported during non-feeding activities), which includes stylet penetration into the diet without ingestion. Those three waveforms occur consecutively and were only recorded on artificial diet (Hori and Naito 2018). Waveform Is represents the stylet insertion into the cotton diet pad. During NI the stylets are motionless within the diet, but without ingestion activity, and W represents the stylet withdrawal from the cotton diet pad (Hori and Naito 2018).

5.4.4 Comparison of EPG Waveforms Among Mirid Genera

The species *L. lineolaris* and *L. hesperus* represent the mirids most extensively studied with EPG, including the first published paper for this family (Cline and Backus 2002) and detailed waveform characterization and description. As mirid waveforms appear to share strong similarities, EPG analyses for the remaining four species (*A. lucorum*, *N. tenuis*, *S. rubrovittatus* and *T. caelestialium*) relied on previous research with *Lygus* spp. (Cline and Backus 2002; Cervantes et al. 2016). Therefore, these studies emphasized quantitative comparisons, rather than detailed waveform description and characterization.

In summary, the feeding behavior of mirids on plants/insect prey can be grouped into five main phases: (1) non-feeding, (2) probing, (3) cell rupturing, (4) ingestion, and (5) transition. Non-feeding includes waveforms that represent any activity where the stylets are not inserted into the food tissue, for example, walking, standing still, or touching the plant surface with antennal tips. This last behavior was only reported for *Lygus* spp. (Cervantes et al. 2016); the lack of a specifically named waveform in other studies might in part be explained by similarities with waveforms correlated with walking activities, making detection difficult. Probing, which is defined here as brief stylet penetration into the substrate, has been reported in *L. hesperus* (Cline and Backus 2002), *A. lucorum* (Li et al. 2016), *T. caelestialium*, and *S. rubrovittatus* (Suzuki and Hori 2014). Cervantes et al. (2016) did not report this behavior separately, but suggested it as a lone cell rupture event for *Lygus* spp.

The cell rupture behavior, previously called drilling/laceration behavior (Cline and Backus 2002), represents degradation of the tissue, mostly via maceration. Ingestion of the solubilized food tissue follows. These two behaviors are the main stylet feeding behaviors performed and were reported in all species of mirids feeding on plants and/or prey insects as well as artificial diet. On plants, the tissue exploited as food source depends on the host, developmental stage, plant structure, etc. Based on histological studies, Cervantes et al. (2017) observed that cell rupture and ingestion waveforms from *L. lineolaris* occurred in the staminal column of the pinhead cotton square. In contrast, for *N. tenuis*, these waveforms occurred in the vascular ring region of the tomato petiole (Chinchilla-Ramírez et al. 2021).

The transition waveform (interpreted as X waves) was only reported for *Lygus* spp. (Cervantes et al. 2016) and *N. tenuis* (Chinchilla-Ramírez et al. 2021). This waveform is clearest when recorded at an input impedance of 10^9 Ohms (Cervantes et al. 2016), which probably explains why Cline and Backus (2002), using an AC monitor at 10^6 Ohms, did not observe it. This transitional waveform was rarely observed and not clearly distinguishable for *N. tenuis* (Chinchilla-Ramírez et al. 2021), and it was absent (or at least not separated) for *A. lucorum* (Li et al. 2016), even though both studies applied Ri of 10^9 Ohms. Perhaps, the difficulty in distinguishing this kind of waveform could be caused by the DC signal applied (both studies used the DC monitor). For *T. caelestialium* and *S. rubrovittatus*, this transition waveform also was not reported, but the authors used a different type of DC-EPG machine and no information about Ri is given.

Finally, the so-called resting behavior was often observed for *T. caelestialium* and *S. rubrovittatus*, but only when they were offered a moistened cotton pad containing a substance to be tested (Hori and Naito 2018). During "resting behavior," the stylets are inserted into this artificial diet, but no ingestion occurs. Cline and Backus (2002) did not report this type of behavior when *L. hesperus* fed on artificial diet, but these authors used a container full of liquid diet, instead of a moistened cotton pad as food source. Perhaps, ingestion behavior was inhibited by the presence of the cotton, or the insects failed to be stimulated by the food when inserting the stylets into the diet containing a specific substance (Hori and Naito 2018).

5.5 EPG Research on Plataspidae

Plataspidae is closely related to Pentatomidae. Feeding behavior of only one species has been determined via EPG, the kudzu bug, *Megacopta cribraria* (F.). This true bug is native to India and Asia and was recently introduced in the USA, where it spread quickly (Gardner 2016). *M. cribraria* is phytophagous, feeding on a large number of wild and cultivated plant species (over 100 associated plants), including several species of Fabaceae (Eger et al. 2018). Since its introduction in the USA, the bug became established as a pest of soybean, on which it prefers vegetative structures, especially stems (Seiter et al. 2013). In addition to its agricultural impact, *M. cribraria* is a home invader causing nuisance problems in urban areas (Ruberson et al. 2013).

The feeding behavior of *M. cribraria* adult females was recorded on soybean stems using an early model of the AC-DC monitor. Six different waveform types were generated, half representing non-feeding (Z1, Z2, and Z3) and half representing feeding activities (G, H, and I) (Stubbins et al. 2017). These waveforms and their proposed biological meanings are summarized in Table 5.5.

Among the non-feeding behavior waveforms, Z1 is marked by irregular peaks with variable amplitude and visually correlates with the bug walking and/or grooming on the stem surface. Z2 is a waveform with only slight variation in amplitude (considered the baseline of recording) and correlated with the bug resting on a

Table 5.5 Summary of the EPG waveforms of *Megacopta cribraria* adult females recorded on soybean stems and their proposed biological activities

Phase	Family	Type	Proposed biological activities
Non-feeding	Z	Z1	Walking, grooming
		Z2	Standing still
		Z3	Mouthpart dabbing
Pathway	G	–	Mouthparts entering plant tissue, salivary sheath formation
	H	–	Transition into/out of ingestion tissue
Ingestion	I	–	Ingestion from xylem/phloem

Source: adapted from Stubbins et al. (2017)

soybean stem. Z3 is characterized by regular peaks; it is thought to represent labial contact with the stem surface before stylet insertion (Stubbins et al. 2017). Z3 is similar to waveform R recorded for *Piezodorus guildinii* (Westwood) on soybean, which is also observed before stylets are inserted (Lucini et al. 2016).

Two distinct pathway waveforms were observed during feeding behavior of *M. cribraria* on soybean stem, G and H. Waveform G is the first feeding waveform and its stereotypical pattern is composed of a set of episodes, each containing peaks. This wave has high voltage level at the beginning which decreases over time. Stubbins et al. (2017) propose that type G represents stylet penetration and secretion of the salivary sheath while stylets are being inserted deeply in the tissue. At the end of feeding, a reverse, short duration version of G waveform is recorded, representing stylet removal from plant tissue (Stubbins et al. 2017).

In sequence, a second pathway type (H) follows the initial waveform G. Waveform H is irregular, but often containing short episodes of high, regular spikes. It resembles waveform D from the psyllid *Diaphorina citri* Kuwayama (Bonani et al. 2010), which is correlated with salivary sheath tip positioned in an unknown phloem tissue and is considered a transition behavior between pathway and phloem activities. A single histological section during waveform H shows the salivary sheath terminating in the non-conductive phloem fiber cap. Therefore, waveform H probably represents this transition behavior before the kudzu bug reaches the conductive phloem tissue (Stubbins et al. 2017).

Ingestion phase (waveform I) follows waveform H and presents a regular pattern, although transitional waveforms are also recorded (Stubbins et al. 2017). The salivary sheath was usually positioned in phloem tissue, but one section shows a sheath tip ending in the xylem vessels. Therefore, Stubbins et al. (2017) associated waveform I with ingestion from both vascular tissues, but primarily in phloem. Seiter et al. (2013) observed presence of fungal colonies growing on liquids (honeydew) secreted by *M. cribraria*, also indicating ingestion from phloem cells.

5.6 Comparison of EPG Waveforms Among Heteropteran Families

Non-feeding activities performed by heteropterans include a variety of behaviors, each one represented by one or more waveforms. The most common non-feeding activities are standing still and walking on the plant surface. Each one is associated with a characteristic waveform, similar in appearance among true bug species. Other non-feeding activities include antennation, labial dabbing, egg laying, and excretion. Although heteropterans typically explore the plant surface with the antennae (Backus 1988), a specific antennation waveform has been reported only for *Lygus* bugs; it is composed of single or multiple spikes that represent touching of the antennal tips to the plant surface (Cervantes et al. 2016). It is likely that in EPG studies of other bugs, antennation behavior has simply been subsumed into the "walking" waveform, which was noted by several authors to include antennation, as well as other activities such as grooming and clawing (Maskey 2010; Stubbins et al. 2017; Mitchell et al. 2018). The antennation spike, being entirely of emf origin (Cervantes et al. 2016), does not register at low input impedances, which may have contributed to its conflation with other non-feeding waveforms.

Labial dabbing, in which the bug touches the plant surface with the labium before stylet insertion, was observed visually in a number of species as one of the behaviors included in a non-feeding waveform, including *E. meditabunda* (Lucini and Panizzi 2016), *H. halys* (Serteyn et al. 2020), and *A. tristis* (Maskey 2010). However, it has been identified as a specific waveform in four species: *A. tristis*, *P. guildinii*, *M. cribraria*, and *N. viridula* (Cook and Neal 1999; Lucini et al. 2016; Stubbins et al. 2017; Mitchell et al. 2018). Waveforms that represent labial contact (R in *P. guildinii*, Z3 in *M. cribraria* and *N. viridula*) are similar among these species.

Egg laying and regurgitation/re-ingestion were each recorded in a single pentatomid species (Lucini and Panizzi 2017a, 2018b, respectively), although similar re-ingestion from a droplet was observed (but not recorded) in another species (Lucini and Panizzi 2017b) (see Chap. 8 for illustrations and further details of these behaviors). Furthermore, nymphs of *N. viridula* are known to imbibe directly from drops of water (Lockwood and Story 1986; Rivera and Mitchell, unpublished); thus, ingestion by bugs from some form of liquid droplets may be a more common behavior than is generally realized. However, the apparent rarity of regurgitation behavior following xylem ingestion is surprising, considering the large number of species that hydrate from xylem, and deserves further study. Another excretion activity was reported for *L. hesperus*; this very short behavior is related to anal drop production while nymphs fed on plants (Cline and Backus 2002). Using a modern EPG monitor, Cervantes et al. (2016) did not report this behavior; but they did not use video recordings to correlate the waveform as was done in the earlier study, thus, this short excretion waveform might have passed unnoticed.

Pathway activities are reported for all true bug species studied with EPG except Miridae (Cimicomorpha). The pathway phase represents secretion of gelling saliva to form the trunk and branches of the salivary sheath, which surrounds the stylets as they penetrate into tissue (Backus et al. 2013; Lucini and Panizzi 2018a; Stubbins et al. 2017); only Pentatomomorpha are known to secrete such a sheath (Miles 1972). Pathway phase comprises a variable number of waveform types/subtypes, each one correlated with a different behavior. However, each waveform that represents a specific stylet activity generally shares a similar appearance (mostly highly jagged waveforms) and electrical characteristics across species.

Mirids perform an array of feeding behaviors that culminates in the degradation of the plant tissue using a cell rupturing strategy (Miles 1972; Backus et al. 2007; Cervantes et al. 2016) without producing gelling saliva. Although a number of pentatomid species also have been reported to use the cell rupture strategy, the tactic employed is different. Pentatomids follow a pathway phase, generally secreting an incomplete salivary sheath in the plant tissue (Lucini and Panizzi 2018a).

Lacerating and/or macerating waveform(s), as tactics of the cell rupturing strategy, have been recorded for both pentatomids and mirids and in part are similar among them. In mirids, the stylets are held relatively still during cell rupture; the slow movement performed by the insects is thought to spread watery saliva for macerating. In contrast, in pentatomids the stylets are moved extensively up and down into the tissue to lacerate and also spread watery saliva. These activities produce waveforms with irregular (lacking discernible patterns) and variable appearance and amplitude among species, between individual insects, and even within the same insect recording. However, as the cell rupture waveform of mirids involves slow stylet movements, the waveform is a mixture of flat sections (very low frequency) randomly interspersed with sections of variable, sometimes high, frequency. In contrast, in pentatomids the continuous and deep stylet movements promoted by laceration activities cause abrupt voltage changes and waveforms primarily displaying a higher frequency; i.e., flat sections are rarely observed. Perhaps, the higher frequency sections observed during cell rupturing by mirids relate to the moments that stylets are being moved, even though slowly.

The waveform related to ingestion of degraded cell contents is similar among mirids and pentatomids; it consists of a low amplitude (at low Ri levels), highly regular pattern, sometimes separated into peak (up or down oriented) and wave portions, with a characteristic frequency. However, presence of peaks and their amplitude and orientation varied greatly with species and Ri level applied (Cervantes et al. 2016; Lucini and Panizzi 2018a). In pentatomids, laceration/maceration and ingestion activities occur interspersed with each other throughout the entire feeding event; i.e., the bugs perform multiple short events of each activity (Lucini and Panizzi 2018a). In mirids, these distinct activities do not occur interspersed during a feeding event; instead, the bugs first break up cells via laceration and/or maceration, then test for the extent of cell degradation, and finally proceed to ingest (Cervantes et al. 2016).

The ingestion period in pentatomids is always much shorter than the time devoted to laceration/maceration activities. For example, of the total time spent by

Euschistus heros (F.) in cell rupture activities on soybean seeds, almost 97% represents lacerating/macerating activities and only 3% ingestion (Lucini and Panizzi 2018b). In contrast, for *Lygus* spp., ingestion events generally present longer durations, occupying >45% of the total time devoted to cell rupture plus ingestion activities (Cervantes et al. 2016).

Species-specific waveforms (X waves) are reported for blissids, mirids, and pentatomids. These brief transition waveforms are primarily performed by salivary sheath feeders that ingest sap from vascular tissues (e.g., Backus et al. 2009; Lucini and Panizzi 2016; Chuche et al. 2017). For blissids (*B. insularis* and *B. occiduus*) and pentatomids (*E. meditabunda* and *H. halys*), X waves are reported preceding phloem ingestion (Backus et al. 2013; Lucini and Panizzi 2016; Serteyn et al. 2020). The phloem X wave recorded for *E. meditabunda* strongly resembles the X wave recorded for *Blissus* spp. when an input impedance of 10^9 Ohms is applied. For both insect species, the X wave appears as a highly regular and long underlying waveform with a gradual increase of voltage level towards the end. A brief and abrupt voltage drop or increase marks the end of the X wave and the beginning of sustained phloem ingestion. In contrast, the fine structure of the putative phloem X wave from *H. halys* does not resemble those recorded for chinch bugs and stink bugs. Although most true bugs have been recorded hydrating from xylem (Backus et al. 2013; Lucini and Panizzi 2018a), only *H. halys* is reported to produce an X wave preceding xylem ingestion (Serteyn et al. 2020).

An X wave is associated with cell rupture feeding only in *Lygus* spp. This transition waveform is composed of repetitive episodes, each one containing three subtypes; the first and second subtypes present a regular pattern, whereas the third subtype is irregular. Each subtype represents specific stylet activities in the plant and strongly resembles some components of other X waves recorded, especially those of sharpshooters. Thus, they probably share the same biological meanings (Cervantes et al. 2016). In salivary sheath feeders, X waves are related to testing and tasting vascular cells for acceptance or rejection (Backus et al. 2009). In contrast, for *Lygus* spp. waveform X probably represents testing and tasting of macerated plant tissue to achieve the greatest level of liquefaction before ingestion (Cervantes et al. 2016).

Multiple pentatomid species perform cell rupture on their hosts, but no X wave has been reported preceding ingestion events. Perhaps, due to the fact that mirids use only the maceration tactic during feeding, the enzymatic degradation of the cells might take time to achieve a proper slurry consistency and thus requires tasting before ingestion. In contrast, pentatomids seem to mix laceration and maceration tactics to potentialize and accelerate the cell degradation, so that tasting the slurry might not be required before ingestion. Furthermore, the ingestion event in mirids is of long duration. In pentatomids, it is, in general, very short (a few seconds) and repeated several times over the course of the feeding activity, in which case tasting would make little sense and likely would not be necessary.

With the exception of mirids (Cimicomorpha), all other true bugs (Pentatomomorpha – blissids, coreids, pentatomids, and plataspids) evaluated with EPG ingest sap from either xylem or phloem and sometimes from both tissues

(Bonjour et al. 1991; Backus et al. 2013; Lucini and Panizzi 2018a,b; Stubbins et al. 2017). However, no heteropteran relies solely on xylem ingestion for nutrition; ingestion from an alternate tissue (phloem, endosperm, or parenchyma) is the primary source of nutrients. Xylem is thought to provide mainly hydration (Lucini and Panizzi 2018a; Rivera and Mitchell 2020). Waveforms related to ingestion from xylem share strong similarities (shape and electrical characteristics) among true bugs tested. Xylem ingestion is characterized by a stereotypical waveform showing a very regular pattern composed of repetitive episodes; each one contains a wave portion interspersed with peaks either positive- or negative-going.

Phloem ingestion was reported for blissids, two pentatomids (*E. meditabunda* and *H. halys*) and a plataspid. In summary, the *E. meditabunda* phloem waveform shows a stereotypical sinuous pattern with multiple episodes of down and up; the underlying waves are very regular, without peaks. In contrast, in the remaining true bug species, this sinuous pattern was not reported; in addition, the wave portions of the phloem waveforms are interspersed with multiple peaks. Despite these differences in appearance, the electrical characteristics of the waveform are very similar among phloem-ingesting species. In general, it is a higher regular frequency waveform with lower relative amplitude compared to xylem ingestion and presents longer event durations.

Sometimes, very short interruption waveforms are recorded during ingestion from vascular tissues. These types of waveforms are reported primarily during xylem ingestion, on blissids and some species of pentatomids, and are similar among them. These interruptions are, in general, characterized by a positive flat-spiky plateau. Interruptions during xylem ingestion are related to testing of xylem cells and always observed at the beginning of the ingestion event (Backus et al. 2013; Lucini and Panizzi 2016, 2018b; Lucini et al. 2016), whereas, in the phloem (only observed in the pentatomid *E. meditabunda*), interruptions are observed throughout the entire ingestion period. Chinch bugs do not present interruptions during phloem ingestion, but Backus et al. (2013) indicate a single waveform that probably represents salivation behavior in phloem cells, which is recorded at the start of phloem phase, as observed in aphids. No waveform like this was observed in *E. meditabunda*; thus, the multiple interruptions might be the sole salivation behavior in phloem cells. In fact, *E. meditabunda* is so far the only true bug to present interruptions during both phloem and xylem ingestion (Lucini and Panizzi 2016).

5.7 Concluding Remarks

In this chapter we have presented and discussed in detail the feeding activities of true bugs (other than stink bugs) analyzed using electronic monitoring and compared waveforms across the two infraorders and all five families thus far examined using this technique. Interestingly, species within the same infraorder (Pentatomomorpha) present similar activities across several families, whereas phytophagous bugs from another infraorder (Cimicomorpha) show some distinctly different feeding activities. Although previous studies based on visual observations or video recordings have also shown differences in feeding behavior, the waveforms generated by EPG are revealing details of the feeding process unknown until now, providing a turning point in our knowledge of this subject.

References

Backus EA (1988) Sensory systems and behaviours which mediate hemipteran plant feeding: a taxonomic overview. J Insect Physiol 34:151–165

Backus EA, Bennett WH (1992) New AC electronic insect feeding monitor for fine-structure analysis of waveforms. Ann Entomol Soc Am 85:437–444

Backus EA, Habibi J, Yan F, Ellersieck M (2005) Stylet penetration by adult *Homalodisca coagulata* on grape: electrical penetration graph waveform characterization, tissue correlation, and possible implications for transmission of *Xylella fastidiosa*. Ann Entomol Soc Am 98:787–813

Backus EA, Cline AR, Ellerseick MR, Serrano MS (2007) *Lygus hesperus* (Hemiptera: Miridae) feeding on cotton: new methods and parameters for analysis of nonsequential electrical penetration graph data. Ann Entomol Soc Am 100:296–310

Backus EA, Holmes WJ, Schreiber F, Reardon BJ, Walker GP (2009) Sharpshooter X wave: correlation of an electrical penetration graph waveform with xylem penetration supports a hypothesized mechanism for *Xylella fastidiosa* inoculation. Ann Entomol Soc Am 102:847–867

Backus EA, Andrews KB, Shugart HJ, Greve LC, Labavitch JM, Alhaddad H (2012) Salivary enzymes are injected into xylem by the glassy-winged sharpshooter, a vector of *Xylella fastidiosa*. J Insect Physiol 58:949–959

Backus EA, Rangasamy M, Stamm M, McAuslane HJ, Cherry R (2013) Waveform library for chinch bugs (Hemiptera: Heteroptera: Blissidae): Characterization of electrical penetration graph waveforms at multiple input impedances. Ann Entomol Soc Am 106:524–539

Backus EA, Cervantes FA, Guedes RNC, Li AY, Wayadande AC (2019) AC-DC electropenetrography for in-depth studies of feeding and oviposition behaviors. Ann Entomol Soc Am 112:236–248

Baxendale FP, Heng-Moss TM, Riordan TP (1999) *Blissus occiduus* (Hemiptera: Lygaeidae): a chinch bug pest new to buffalograss turf. J Econ Entomol 92:1172–1176

Bonani JP, Fereres A, Garzo E, Miranda MP, Appezzato-da-Gloria B, Lopes JRS (2010) Characterization of electrical penetration graphs of the Asian citrus psyllid, *Diaphorina citri*, in sweet orange seedlings. Entomol Exp Appl 134:35–49

Bonjour EL, Fargo WS, Webster JA, Richardson PE, Brusewitz GH (1991) Probing behavior comparisons of squash bugs (Heteroptera: Coreidae) on cucurbit hosts. Environ Entomol 20:143–149

Calderon JD, Backus EA (1992) Comparison of the probing behaviors of *Empoasca fabae* and *E. kraemeri* (Homoptera: Cicadellidae) on resistant and susceptible cultivars of common beans. J Econ Entomol 85:88–99

Cervantes FA, Backus EA, Godfrey L, Akbar W, Clark TL (2016) Characterization of an EPG waveform library for pre-reproductive adult *Lygus lineolaris* and *L. hesperus* feeding on cotton squares. Ann Entomol Soc Am 109:684–697

Cervantes FA, Backus EA, Godfrey L, Wallis C, Akbar W, Clark TL, Rojas MG (2017) Correlation of electropenetrography waveforms from *Lygus lineolaris* (Hemiptera: Miridae) feeding on cotton squares with chemical evidence of inducible tannins. J Econ Entomol 110:2068–2075

Chinchilla-Ramírez M, Garzo E, Fereres A, Gavara-Vidal J, ten Broeke CJM, van Loon JJA, Urbaneja A, Pérez-Hedo M (2021) Plant feeding by *Nesidiocoris tenuis*: quantifying its behavioral and mechanical components. Biol Control 152:104402;1–10

Chuche J, Backus EA, Thiéry D, Sauvion N (2017) First finding of a dual-meaning X wave for phloem and xylem fluid ingestion: characterization of *Scaphoideus titanus* (Hemiptera: Cicadellidae) EPG waveforms. J Insect Physiol 102:50–61

Cline AR, Backus EA (2002) Correlations among AC electronic monitoring waveforms, body postures, and stylet penetration behaviors of *Lygus hesperus* (Hemiptera: Miridae). Environ Entomol 31:538–549

Cook CA, Neal JJ (1999) Feeding behavior of larvae of *Anasa tristis* (Heteroptera: Coreidae) on pumpkin and cucumber. Environ Entomol 28:173–177

De Puysseleyr V, De Man S, Höfte M, De Clercq P (2013) Plantless rearing of the zoophytophagous bug *Nesidiocoris tenuis*. BioControl 58:205–213

Eger JE, Gardner WA, Greene JK, Jenkins TM, Roberts PM, Suiter DR (2018) *Megacopta cribraria* (F.). In: McPherson JE (ed) Invasive stink bugs and related species (Pentatomoidea): Biology, higher systematics, semiochemistry, and management. CRC Press, Boca Raton, pp 293–332

Eickhoff TE, Baxendale FP, Heng-Moss TM, Blankenship EE (2004) Turfgrass, crop, and weed hosts of *Blissus occiduus* (Hemiptera: Lygaeidae). J Econ Entomol 97:67–73

Gardner WA (2016) Kudzu bug distribution map. University of Georgia Center for Invasive Species and Ecosystem Health, Athens, GA. Available in: www.kudzubug.org. Accessed 05 June 2020

Hori K (2000) Possible causes of disease symptoms resulting from the feeding of phytophagous Heteroptera. In: Schaefer CW, Panizzi AR (eds) Heteroptera of economic importance. CRC Press, Boca Raton, pp 11–36

Hori M, Naito S (2018) Feeding behaviors of rice ear bugs, *Trigonotylus caelestialium* and *Stenotus rubrovittatus* (Hemiptera: Miridae), in response to starch and its related substances. Appl Entomol Zool 53:143–150

Jin S, Chen ZM, Backus EA, Sun XL, Xiao B (2012) Characterization of EPG waveforms for the tea green leafhopper, *Empoasca vitis* Göthe (Hemiptera: Cicadellidae), on tea plants and their correlation with stylet activities. J Insect Physiol 58:1235–1244

Li W, Wyckhuys KAG, Wu K (2016) Does feeding behavior of a zoophytophagous mirid differ between host plant and insect prey items? Arthropod Plant Interact 10:79–86

Lockwood JA, Story RN (1986) Adaptive functions of nymphal aggregation in the southern green stink bug, *Nezara viridula* (L.) (Hemiptera: Pentatomidae). Environ Entomol 15:739–749

Lu Z, Dong S, Li C, Li L, Yu Y, Men X, Yin S (2020) Sublethal and transgenerational effects of dinotefuran on biological parameters and behavioural traits of the mirid bug *Apolygus lucorum*. Sci Rep 10:226;1–8

Lucini T, Panizzi AR (2016) Waveform characterization of the soybean stem feeder *Edessa meditabunda* (F.) (Hemiptera: Heteroptera: Pentatomidae): overcoming the challenge of wiring pentatomids for EPG. Entomol Exp Appl 158:118–132

Lucini T, Panizzi AR (2017a) Feeding behavior of the stink bug *Dichelops melacanthus* Dallas on corn seedlings: an EPG analysis at multiple input impedances and histology correlation. Ann Entomol Soc Am 110:160–171

Lucini T, Panizzi AR (2017b) Behavioral comparisons of ingestion and excretion by selected species of pentatomids: evidence of feeding on different food sources supports pest status. Neotrop Entomol 46:361–367

Lucini T, Panizzi AR (2018a) Electropenetrography (EPG): a breakthrough tool unveiling stink bug (Pentatomidae) feeding on plants. Neotrop Entomol 47:6–18

Lucini T, Panizzi AR (2018b) EPG monitoring of the Neotropical brown-stink bug, *Euschistus heros* (F.), on soybean pods: An electrical penetration graph-histology analysis. J Insect Sci 18:5;1–14

Lucini T, Panizzi AR, Backus EA (2016) Characterization of an EPG waveform library for red-banded stink bug, *Piezodorus guildinii* (Hemiptera: Pentatomidae), on soybean plants. Ann Entomol Soc Am 109:198–210

Maskey K (2010) Comparison of electrical penetration graph waveforms of squash bug feeding on watermelon and its relatives. Master's thesis, Oklahoma State University, Stillwater, Oklahoma

Miles PW (1972) The saliva of Hemiptera. Adv Ins Physiol 9:183–255

Miles PW, Taylor GS (1994) 'Osmotic pump' feeding by coreids. Entomol Exp Appl 73:163–173

Mitchell PL (2000) Leaf-footed bugs (Coreidae). In: Schaefer CW, Panizzi AR (eds) Heteroptera of economic importance. CRC Press, Boca Raton, pp 337–403

Mitchell PL (2006) Polyphagy in true bugs: a case study of *Leptoglossus phyllopus* (L.) (Hemiptera, Heteroptera, Coreidae). Denisia 19, Neue Serie 50:1117–1134

Mitchell PL, Cooke SB, Smaniotto LF (2018) Probing behavior of *Nezara viridula* on soybean: characterization and comparison of electrical penetration graph (EPG) waveforms on vegetative and reproductive plant structures. J Agric Urban Entomol 34:19–43

Neal JJ (1993) Xylem transport interruption by *Anasa tristis* feeding causes *Cucurbita pepo* to wilt. Entomol Exp Appl 69:195–200

Prado E, Tjallingii WF (1994) Aphid activities during sieve element punctures. Entomol Exp Appl 72:157–165

Reinert JA, Kerr SH (1973) Bionomics and control of lawn chinch bugs. Bull Entomol Soc Am 19:91–92

Rivera KL, Mitchell PL (2020) Probing behavior of *Nezara viridula* first instars: EPG analysis and effect of food availability on subsequent development. J Agric Urban Entomol 36:47–63

Rodrigues D, Sampaio DS, Isaias RMDS, Moreira GRP (2007) Xylem and seed feeding by two passion vine leaffooted bugs, *Holymenia clavigera* and *Anisoscelis foliacea marginella* (Hemiptera: Coreidae: Anisoscelini) with notes on mouthpart morphology. Ann Entomol Soc Am 100:907–913

Ruberson JR, Takasu K, Buntin GD, Eger JE Jr, Gardner WA, Greene JK, Jenkins TM, Jones WA, Olson DM, Roberts PM, Suiter DR, Toews MD (2013) From Asian curiosity to eruptive American pest: *Megacopta cribraria* (Hemiptera: Plataspidae) and prospects for its biological control. Appl Entomol Zool 48:3–13

Schaefer CW, Mitchell PL (1983) Food plants of the Coreoidea (Hemiptera: Heteroptera). Ann Entomol Soc Am 76:591–615

Seiter NJ, Greene JK, Reay-Jones FPF (2013) Reduction of soybean yield components by *Megacopta cribraria* (Hemiptera: Plataspidae). J Econ Entomol 106:1676–1683

Serteyn L, Ponnet L, Backus EA, Francis F (2020) Characterization of electropenetrography waveforms for the invasive heteropteran pest, *Halyomorpha halys*, on *Vicia faba* leaves. Arthropod Plant Interact 14:113–126

Stubbins FL, Mitchell PL, Turnbull MW, Reay-Jones FPF, Greene JK (2017) Mouthpart morphology and feeding behavior of the invasive kudzu bug, *Megacopta cribraria* (Hemiptera: Plataspidae). Invertebr Biol 136:309–320

Suzuki Y, Hori M (2014) Diurnal locomotion and feeding activities of two rice-ear bugs, *Trigonotylus caelestialium* and *Stenotus rubrovittatus* (Hemiptera: Heteroptera: Miridae). Appl Entomol Zool 49:149–157

Tjallingii WF (1988) Electrical recording of stylet penetration activities. In: Minks AK, Harrewjin P (eds) Aphids: their biology, natural enemies and control. Elsevier, Amsterdam, pp 95–108

Walker GP, Medina-Ortega HJ (2012) Penetration of faba bean sieve elements by pea aphid does not trigger forisome dispersal. Entomol Exp Appl 144:326–335

Wheeler AG (2000a) Plant bugs (Miridae) as plant pests. In: Schaefer CW, Panizzi AR (eds) Heteroptera of economic importance. CRC Press, Boca Raton, pp 37–83

Wheeler AG (2000b) Predacious plant bugs (Miridae). In: Schaefer CW, Panizzi AR (eds) Heteroptera of economic importance. CRC Press, Boca Raton, pp 657–693

Yuan W, Li WJ, Lu YH, Wu KM (2013) Combination of plant and insect eggs as food sources facilitates ovarian development in an omnivorous bug *Apolygus lucorum* (Hemiptera: Miridae). J Econ Entomol 106:1200–1208

Chapter 6
EPG Procedures for True Bugs (Heteroptera)

Contents

Abstract Since its development, the electrical penetration graph (EPG) has been used to record the feeding behavior of mostly small, relatively weak insects (aphids and leafhoppers). Recently the larger, more robust true bugs (e.g., pentatomids) began to be monitored via EPG, and some problems were encountered, especially related to gold wire attachment. Wiring procedures involve the choice of an electrically conductive adhesive (silver paint or silver glue) and methods to immobilize the bugs to facilitate the process and guarantee successful wiring by sanding the cuticle. Histological studies are often coupled with EPG recordings to correlate waveforms with the location of the salivary sheaths or stylets in the plant tissue. EPG can also be applied in quantitative studies to determine counts and durations of the waveforms recorded and then to compare via statistical analysis. At the start of EPG research, essential variables must be standardized, such as the ideal starvation period (if required), recording times, and laboratory environmental conditions. Appropriate experimental design and the acceptable sample size (number of insects recorded per treatment) are also important considerations.

Keywords Wiring procedures · Data analysis · Waveform correlation · EPG variables · Standardization

A. R. Panizzi et al., *Electronic Monitoring of Feeding Behavior of Phytophagous True Bugs (Heteroptera)*, Entomology in Focus 6,
https://doi.org/10.1007/978-3-030-64674-5_6

6.1 Introduction

In this chapter, we will highlight and discuss some points related to electrical penetration graph (EPG) procedures for true bugs, including wiring protocols, data analyses, waveform correlation, etc. Over time, the use of EPG with true bugs has revealed some problems and limitations that needed to be overcome. True bugs are, in general, large, vigorous, and more active compared to the smaller insects commonly recorded via EPG (e.g., aphids). Such characteristics directly affect the wire attachment, as reported for stink bugs (Lucini and Panizzi 2016). Early EPG research with true bugs used a wiring process similar to that of other sucking insects but recorded only small heteropteran species (e.g., Blissidae) or nymphs of larger ones (e.g., Coreidae). However, for large and restless true bugs, the method needed improvement to guarantee a successful attachment. Lucini and Panizzi (2016) developed a sanding protocol to improve wiring success in stink bugs, which might be applicable to other vigorous true bug species as well.

In addition to qualitative studies to determine feeding behavior, EPG recordings can be quantified to determine counts and durations of waveforms performed by hemipterans and then statistically compared (Backus et al. 2007). Moreover, the AC-DC EPG monitor permits the potential effects of various electrical settings on feeding behavior to be assessed (i.e., applied signal, input impedance, and voltage level) (Backus et al. 2018; Cervantes and Backus 2018). This knowledge allows researchers to judge and define the best monitor settings for future studies.

EPG recordings are often accompanied by methods to correlate waveforms with feeding behaviors and to identify the plant tissue associated with each waveform. Histological studies are the most useful method to determine the location of the stylets and/or salivary sheath in the plant tissue. Another method is artificial diet coupled with video and EPG recordings simultaneously; stylet activities (e.g., salivation and ingestion) are directly observed and correlated with their respective EPG waveforms. Some crucial considerations must be taken into account before starting EPG recordings, such as (1) deciding whether a starvation period is required, and if so, for how long; (2) determining the ideal recording time; (3) always using the same conditions and an appropriate experimental design; and (4) recording an acceptable number of insects per treatment.

6.2 Wire Attachment Techniques

The insect electrode, termed "wiring stub," is composed of a piece of copper wire (ca. 2–3 cm long) soldered to a brass nail. At the tip of the copper wire, a thin and flexible piece of gold wire is fixed and glued using a conductive adhesive. This piece of gold wire is variable in length (long enough to allow the bug to move and explore the host surface) and diameter (≈ 0.01–0.1 mm), varying according to the insect studied. In general, for stink bugs the diameter used is 0.1 mm, whereas for mirids a thinner wire, < 0.04 mm, is optimal. A small loop is made at the end of the gold wire to expand the contact surface, thereby improving the attachment and conductivity for larger insects [E.A. Backus unpublished data, cited in Cervantes and Backus (2018)].

The use of EPG to monitor true bugs has revealed several problems that needed to be overcome and limitations on the use of the equipment of which one must be aware. Our expertise on such work, we believe, allow us to share our experience in order to "pave the way" to new users, providing tips that will facilitate the conduction of EPG recordings.

Among the problems, wiring vigorous and restless insects such as true bugs is a big challenge. When we first started wiring pentatomids, the biggest difficulty was keeping the gold wire affixed to the dorsum (pronotum) of the bug. The quality of the waveforms obtained during the recording process is directly affected by this fixation. If the gold wire does not perfectly adhere to the insect's body (indicated by the red arrow in Fig. 6.1a), the waveforms generated are not clear, showing variable disturbances (noise) (Fig. 6.1c and d) compared to the normal, clean waveform (Fig. 6.1b), and must be deleted from the final waveform analysis.

In the sections that follow, we will discuss some aspects related to wiring attachment procedures for true bugs, including conductive adhesives used to affix the gold wire to the insect body, and discuss methods employed to facilitate and improve this process.

6.2.1 Conductive Paint/Glue

Two types of electrically conductive adhesive have been used to tether insects with gold wire in EPG studies: silver paint (a commercial product used primarily in the electronics industry) and handmade silver glue. For small aphids and leafhoppers, McLean and Kinsey (1964) used commercial silver paint (silver powder mixture in a solvent) to attach the gold wire. Later, Tjallingii (1978) replaced this adhesive with a water-based adhesive, produced by mixing white glue with silver powder, because some results indicated negative effects of the solvent on aphids. Recently, G. P. Walker [personal communication, cited in Cervantes and Backus (2018)] replaced the silver powder with silver flakes to improve the conductivity of the adhesive. This water-based adhesive is presently the most commonly used in studies, and it is made by mixing silver flakes, household white glue, and water in a 1:1:1 (Wt:Vol:Vol) ratio.

Cervantes and Backus (2018) tested both type of adhesives to evaluate their effects on the waveform appearance of a sharpshooter species and observed that handmade silver glue provides more waveform fine-structure details compared to commercial silver paint. They suggest that silver glue might be more conductive compared to silver paint. No study like this has been done with true bugs; however, the solvents used in silver paint could be toxic if combined with the cuticle sandpapering protocol for large bugs (see Sect. 6.2.2). Therefore, we recommend the use of silver glue for stink bugs.

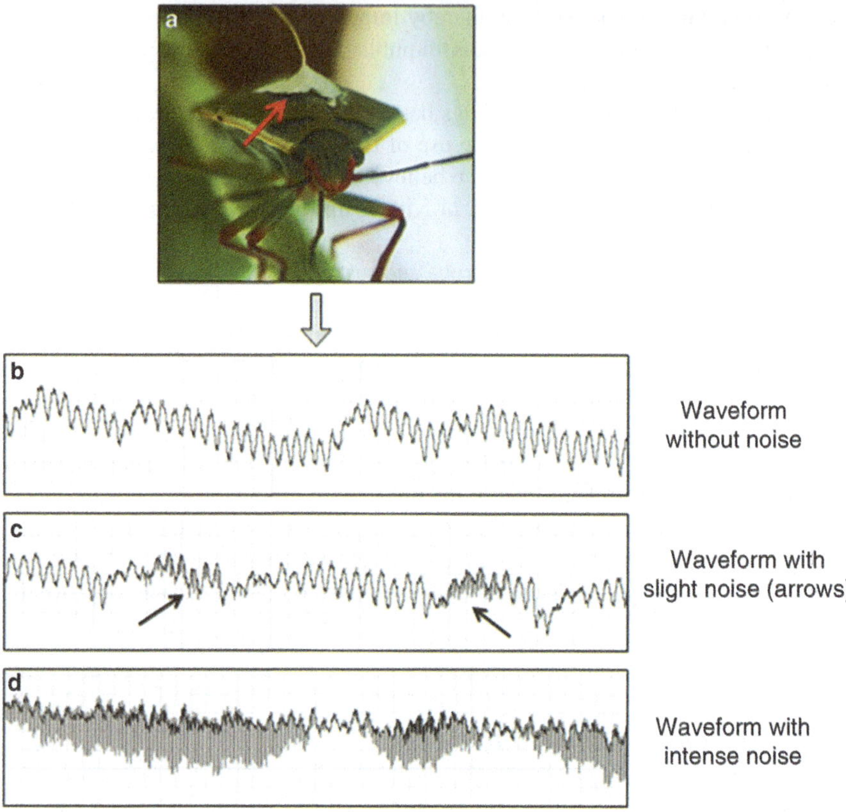

Fig. 6.1 *Chinavia erythrocnemis* (Berg) (Pentatomidae) feeding on soybean stem showing the adhesive not perfectly adhered to the pronotum (loose gold wire indicated by arrow) (**a**). Clean waveform, without noise (**b**). Waveform with slight noises (**c**), and completely overwhelmed by noise (**d**). (Photos by T. Lucini)

6.2.2 Sandpapering the Pronotum

After bugs are wired and placed in position on the EPG, they often pull loose; this is especially problematic with pentatomids. We have developed a method for improving the effectiveness of the gold wire attachment, using human dental sandpaper to remove the cuticular lipids from the gluing area on the pronotum. This creates a scraped area on which a drop of silver adhesive is placed (Fig. 6.2). This sanding and wiring technique (Lucini and Panizzi 2016) also increases the contact area of the silver glue drop: its diameter was significantly greater than the diameter of drops on unsanded bugs. In our studies of wiring pentatomids, we found that bugs with greater mobility (i.e., more restless) have lower wire retention than those that are less mobile, even if larger and heavier (Lucini and Panizzi 2016). Therefore, behavior seems to be the most important trait affecting wire retention in true bugs.

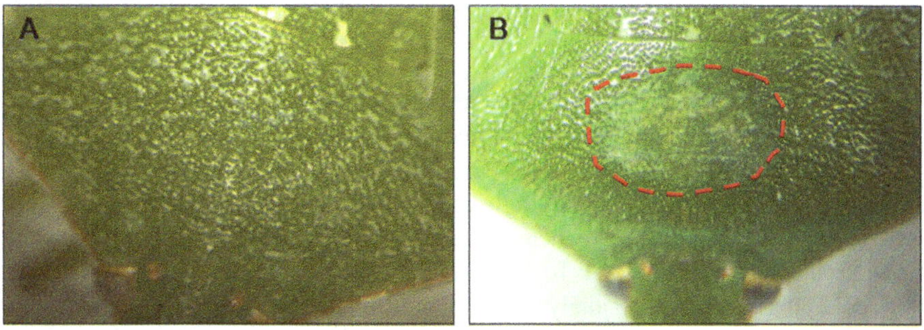

Fig. 6.2 Pronotum of the Southern green stink bug, *Nezara viridula* (L.) before (**a**) and after (**b**) being sanded (area circled in dashed red line), using a piece of dental sandpaper. (Photos by T Lucini)

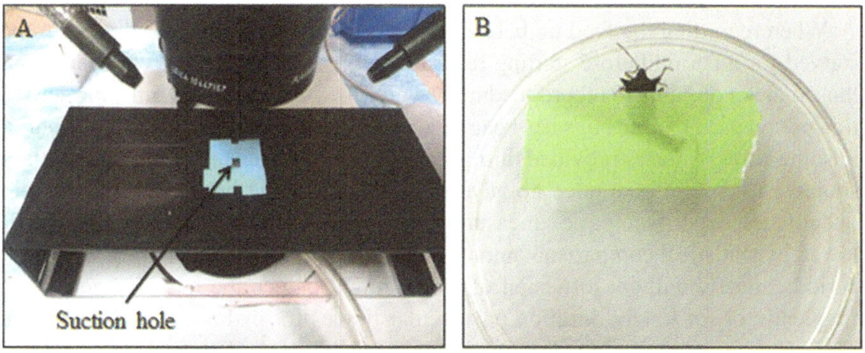

Fig. 6.3 Detail of a vacuum device under a stereomicroscope (**a**). Stink bug held immobile using adhesive tape on the top of a Petri dish lid (**b**). (Photos by T Lucini)

6.2.3 Methods for Restraining Bugs During Wiring

Several techniques have been developed to simplify the wiring procedure. In small and less robust true bugs, such as mirids and blissids, individuals are held immobile by placing the abdomen in a vacuum chamber with low suction pressure under a stereomicroscope (Fig. 6.3a) and finally tethered with the gold wire (Backus et al. 2013; Cervantes et al. 2016; Lu et al. 2020). To make this immobilization an easier process, individuals are previously anesthetized. Insects are enclosed in small vials and then CO_2 gas is applied for a few seconds ($\approx$ 30–60 s) (used for *Lygus* spp. – Cervantes et al. 2016, 2019; Tuelher et al. 2020), or an ice bath is used to chill insects [used for *Apolygus lucorum* (Meyer-Dür), Lu et al. 2020].

For large and more robust true bugs, such as stink bugs, the individuals are held temporarily immobilized by placing their venter on the top of a Petri dish or other surface, in general without anesthesia, and then taping the abdomen with a piece of adhesive tape (labeling tape) leaving the pronotum exposed (Fig. 6.3b). The piece of tape should be rubbed on a cloth before use to reduce stickiness and facilitate removal after the adhesive dries (Mitchell et al. 2018).

6.3 Starvation and Recording Times

For true bugs, different recording access times have been used, ranging from 6 to 24 h uninterrupted. The minimum and maximum access time required to complete feeding behavior varies among bug species. A timeframe that is too short may produce no feeding events, or just a single one that is truncated, making it useless for analysis. Electrical contact may deteriorate over time, so excessively long recording sessions are also to be avoided. To determine a suitable recording time for a species, the researcher must test and analyze a number of preliminary recordings. For example, Cervantes et al. (2016) suggest that 5 h of recording time is sufficient for *Lygus lineolaris* (Palisot de Beauvois) (Miridae) to complete its feeding behavior on one cotton square (bud), because few feeding activities occur after this time. For pentatomids, the majority of the studies used ca. 8 h of plant access time (e.g., Lucini and Panizzi 2018a and references therein).

When recording the feeding behavior of a true bug, it is usual for the bug to be starved for a period before starting recordings, to stimulate the insect to feed and, thus, make it possible to record feeding waveforms. Like recording time, the starvation period used on true bugs has been largely variable (ranging from 2 to 24 hours); in some cases, water is provided during the starvation time. However, some caution should be taken to guarantee no adverse effects on insect behavior, especially in quantitative experiments (i.e., measurement of each waveform event duration followed by statistical comparison among treatments). For example, longer starvation periods, mostly without provision of water, might induce the bug to ingest more frequently or for longer duration on vascular tissues for hydration, mainly xylem. This effect was reported for the pentatomids *Edessa meditabunda* (F.) and *Nezara viridula* (L.) after long starvation periods without water access (Lucini and Panizzi 2016; Mitchell et al. 2018). Likewise, a significant increase in xylem ingestion was observed for other piercing-sucking insects, such as aphids (Spiller et al. 1990) and psyllids (Bonani et al. 2010) when a 24-h starvation period preceded EPG recording. Mitchell et al. (2018) observed that frequency and duration of xylem ingestion were reduced when water was provided to *N. viridula* during the 24-h starvation period. However, the number of 7-h recordings without any ingestion increased significantly under these conditions, so the authors did not recommend provision of a water source (Mitchell et al. 2018). Adjusting and shortening the starvation period would be more effective.

6.4 Data Analysis

Behavioral quantification (waveform counts and durations) performed by hemipterans inside the plant tissue is another powerful contribution of the EPG technique, but this analysis is rather time-consuming. Over time, statistical variables have been proposed to standardize types and terminology. Van Helden and Tjallingii (2000) tried standardizing statistical variables in their analyses, mostly for aphids. They distinguished two classes of EPG variables, sequential (i.e., the sequential order of the waveform events is taken into account) and non-sequential variables (i.e., data are independent of the sequential order of the waveform events) (Table 6.1). However, as aphids present stereotypical sequences during their feeding activities, these authors prioritized the use of sequential variables, which for aphids are more appropriate than non-sequential.

Not all hemipteran species follow a stereotypical, sequential feeding behavior (e.g., non-sheath feeders); therefore, the use of non-sequential response variables may be more convenient than sequential in these cases. A first attempt to organize and designate a complete system primarily focusing on non-sequential variables was presented by Backus et al. (2007). The variables were compiled in a hierarchical scheme which was divided into four levels: cohort, insect, probe, and waveform event; each one containing multiple EPG variables. The terminology "probe" used in relation to EPG variables in this system defines the time from stylet insertion until removal, including all stylet activities performed by the insect (Backus et al. 2007). However, the term probe as stated by these authors has raised controversy (see Preface of this book).

Cohort-level represents the total behavioral information (overall) of the insect population tested (e.g., total waveform duration – TWD). Insect-level represents behavioral information of either each insect individually over the course of the recording or an average across the N insects assayed in the cohort (e.g., waveform duration per insect – WDI). Probe-level combines all waveform events performed from stylet insertion into the plant tissue until removal (e.g., waveform duration per probe per insect – WDPI). Finally, event-level represents a continuous and uninterrupted occurrence of one waveform type in a single probe (e.g., number of waveform events per insect – NWEI) (more details in Backus et al. 2007). Statistical analyses of these non-sequential response variables may be done using the Backus 2.0 SAS program (SAS institute), which was specifically developed to generate and analyze these variables statistically.

Another system to analyze EPG variables is the Sarria program, which is an Excel workbook sheet that analyzes multiple sequential and non-sequential variables. However, it was only designed to summarize results and generates an output data file for subsequent statistical analysis. Furthermore, it was initially developed to calculate EPG variables related to the behavioral activities of aphids

Table 6.1 Examples of sequential and non-sequential EPG variables for aphids

Sequential	Non-sequential
Duration of first np waveform (non-feeding)	Total duration of waveform
Duration of first stylet penetration	Frequency of occurrence of waveform
Time to first E waveform (phloem) from start of recording	Mean duration of waveform
Duration of first E waveform	
Number of penetrations before first E waveform	

Source: adapted from Van Helden and Tjallingii (2000)

(Sarria et al. 2009). Nevertheless, variables in this workbook could be also used for other groups of salivary-sheath feeders, since feeding by these insects presents a behavioral pattern (Backus et al. 2007). The two programs, Backus 2.0 and Sarria workbook, are available to download at https://crec.ifas.ufl.edu/extension/epg/. Five comparative studies with true bugs used the Backus program to obtain and to analyze their EPG variables (Backus et al. 2007, 2018; Cervantes et al. 2019; Lucini and Panizzi 2020; Tuelher et al. 2020). In contrast, only one study used the Sarria workbook to calculate their variables (Chinchilla-Ramírez et al. 2021).

6.5 Monitor Performance

Two EPG monitors have been extensively used to evaluate the behavioral characteristics of hemipterans: 1) the European monitor, Giga-8 model, which uses a fixed input impedance of 10^9 Ohms and DC applied signal (Tjallingii 1988), and 2) the American AC-DC monitor, which allows a choice among multiple input impedance levels (from 10^6 to 10^{10} plus 10^{13} Ohms) and two applied signals, AC and DC (Backus and Bennett 2009). The majority of the true bug species studied via EPG (10 out of 18 species) were evaluated using the AC-DC monitor and only four were evaluated via DC-Giga 8 monitor. An older machine, the Missouri monitor (Backus and Bennett 1992), was used in two studies; this machine represented the last of seven published designs of the original AC monitor but is no longer manufactured. In the remaining studies (*Stenotus rubrovittatus* (Matsumura), *Trigonotylus caelestialium* (Kirkaldy), and early research with *Anasa tristis* (De Geer)) the EPG used was a variant of a DC monitor.

In addition to basic feeding behavior research, EPG can be applied across a wide range of practical studies. These monitors have been used to evaluate and demonstrate the effects of (1) resistant versus susceptible varieties of plants (Rangasamy et al. 2015) and (2) different host and non-host plants (Li et al. 2016; Lucini and Panizzi 2020) on the feeding activities of true bugs. Other agricultural applications of EPG include (3) comparison of behavioral changes caused by the application of pesticides (Lu et al. 2020); (4) insect-induced plant defenses (Serteyn et al. 2020); and (5) comparison of transgenic plants genetically modified to express biopesticides (Cervantes et al. 2019).

The majority of EPG studies with large hemipterans, particularly heteropterans, have been published in the last 10 years, and some questions emerged over time. For example, do electrical signals applied during recordings cause negative effects on the feeding behavior of large insects? During the 50 years of EPG science, it was assumed that the behavioral characteristics of small or tiny insects (aphids, thrips, and psyllids) are not negatively affected by electricity applied during EPG (more details in Backus et al. 2018).

However, anecdotal observations suggested that the use of DC signal and high input impedances may potentially disturb large insects (such as true bugs) during recordings. These monitor settings have been reported to reduce significantly the number and duration of waveform events, especially those related to ingestion activities. However, DC applied signal qualitatively does not change the stylet activities and waveforms performed by the insects; i.e., only quantitative parameters are affected (Backus et al. 2018). Cervantes et al. (2016) observed only slight differences in the appearance of the waveforms recorded for the mirids *Lygus hesperus* Knight and *L. lineolaris* using AC or DC applied signal. In fact, the cell rupturing (CR) and ingestion (I) waveforms were essentially identical with both applied signals. However, they observed that the quantity of feeding behavior could be negatively affected by the type and magnitude of applied voltage used.

The AC-DC monitor allows the real effect of electrical signals on behavior of hemipterans to be evaluated, but few studies have been done. Recently, a robust study (the first for large insects) was carried out using this monitor to test any potential effects of both AC and DC applied signals, and multiple Ri levels and applied voltages, on the feeding behavior of adult *L. lineolaris* (Backus et al. 2018). In summary, the non-feeding and feeding behaviors were directly affected by the signal type used and by Ri level and applied voltage. In general, feeding activity durations were statistically reduced when applied DC signal was used, especially at higher voltage and Ri levels. Based on their results, the authors suggest a combination of AC applied signal, 10^7 Ohms of Ri, and low voltage level (below 50 mV) for future quantitative studies of *L. lineolaris* (Backus et al. 2018).

Another comparative study was performed with the pentatomid, *Bagrada hilaris* (Burmeister) (E. Tulher, T. Lucini, and E. Backus, unpublished data), where the results of AC and DC were directly compared statistically at low and high applied voltage levels. This study demonstrated that DC applied signals and Ri of 10^9 Ohms, coupled with high voltage, directly changed quantification of stink bug waveforms. For stink bugs, preliminary results have demonstrated that the best blend for future quantitative studies is 10^7 Ohms of input impedance and AC applied signal with a low voltage level (50 mV) (Lucini and Panizzi 2018a).

6.6 Standardization of Variables

Some precautions must be taken before recording EPG waveforms of true bugs. For example, recordings must be done always at the same time of the day and under the same conditions (room with continuous light and controlled temperature). It is necessary to determine and standardize the recording time to guarantee the minimum access time for insects to perform all waveforms related to their feeding behavior (see Sect. 6.3). Appropriate experimental design is essential, particularly in quantitative experiments. A randomized block design is recommended; the standard four channels available on both monitor types permit four treatments to be recorded simultaneously and under the same conditions. Furthermore, it is essential to record an adequate number of insects per treatment, because behavioral activities are often variable among individuals. A sample size of at least 15 to 20 insects successfully recorded per treatment is recommended.

Another important point is related to the pretreatment of insects before recordings, which could cause stress and affect behavior; thus one needs to be careful when planning the experiment. Among these pretreatment considerations, we note the wiring process, which includes methods for restraining the bugs (anesthetizing and immobilization) and the starvation period that could directly affect the frequency and duration of ingestion, mostly from xylem tissue, as previously discussed.

For feeding behavior of a true bug species determined for the first time, it is strongly recommended to develop a complete waveform library using multiple input impedances, a relatively low voltage level, and maybe both AC and DC signals, despite preliminary results showing DC signal might cause negative effects. However, an AC-DC monitor is required in order to create a waveform library.

Before starting EPG recording, amplifiers, plants, and insects must be arranged inside a Faraday cage to protect the system against external electrical disturbances (noise), since it is extremely sensitive to any external electrical change (Walker 2000). Researchers might use one large or two small Faraday cages, according to the needs of the studies. Factors to consider include size and developmental stage of plants, whether supports (e.g., helping hands) are required to hold plants or plant parts immobile, and so on.

6.7 Correlating Waveforms with Behaviors

Several methods can be used to correlate waveforms with feeding behaviors. Among these, histological studies have been extensively used in several species of piercing-sucking insects, including true bugs (e.g., Lucini and Panizzi 2016; Stubbins et al. 2017; Mitchell et al. 2018; Almeida et al. 2020). It is the most powerful and accurate method to determine stylet/salivary sheath location within plant tissue, which aids researchers in determining the biological activities of waveforms. Histology allows the entire path taken by the stylets through the plant tissues to be traced, to determine the real position of stylets at the precise time a given waveform is observed. The correlation of the waveform with the specific feeding site is either based on the terminal position of the salivary sheath (Fig. 6.4a) or the terminal position of the severed stylet tips (Fig. 6.4b) during a specific waveform. Salivary sheath position is a reliable indicator because in phytophagous Heteroptera, movement of the mandibular and maxillary stylets is coordinated, and the stylets reach the same depth in the host tissue (see Chap. 2).

Once all waveforms have been characterized for a species, recordings begin again with additional individuals. When a specific waveform is observed on the computer screen, the insect's feeding activities are artificially interrupted or stylets may be severed. After that, the plant tissue bearing the salivary sheath and/or severed stylets is used for the preparation of histological slides. A rotary microtome, which makes fine and precise cuts, may be used for preparation of the slides, but this involves a complex process that requires specific and time-consuming protocols. Hand sections are another option, using a sharp blade (razor blade, bistoury, etc.) or

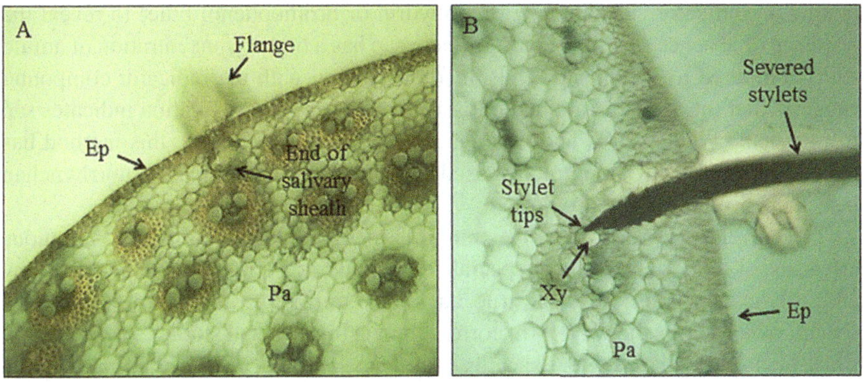

Fig. 6.4 Salivary sheath secreted by the bug *Diceraeus (Dichelops) melacanthus* (Dallas) ending in parenchyma tissues during the pathway waveform (**a**). Stylet tips of the stink bug *Diceraeus (Dichelops) furcatus* (F.) positioned in the xylem cells during their respective waveform. *Ep* epidermis, *Pa* parenchyma, *Xy* xylem. (Source: (**a**) Lucini and Panizzi (2017a); (**b**) photo by T. Lucini)

a hand microtome, which does not require a sophisticated process. Hand sectioning has been done for several species of pentatomids (Lucini and Panizzi 2016, 2017a, b, 2018b; Mitchell et al. 2018; Almeida et al. 2020; Rivera and Mitchell 2020).

Although very useful, the histological studies by themselves cannot indicate the behavior (salivation, ingestion) performed by the insect in the plant tissue, but only the location of the stylet tips. However, other methods may be employed to correlate the waveforms with their behaviors, for example, observations of insects feeding on artificial diet or analysis of excretory droplets. Artificial diet, often containing ink particles, is covered with a parafilm membrane to mimic the plant tissue and placed under a stereomicroscope coupled with a video camera. The diet and the insect are connected to an EPG monitor. Stylet activities, including stylet movements, salivation, and ingestion, may be correlated with their respective waveforms recorded simultaneously; movement of ink particles allows salivation and ingestion to be distinguished.

This method has been used with several piercing-sucking insects (Kindt et al. 2003; Joost et al. 2006; Jin et al. 2012). For true bugs, it was used by Cline and Backus (2002) to correlate the stylet activities of the mirid *L. hesperus* with waveforms and their biological meanings. However, extrapolating results obtained from insects feeding on artificial diet to insects feeding on plant tissues must be done with caution. Direct observations of the stylets in plants that have relatively transparent tissues can supplement studies using artificial diet (Walker 2000).

Another useful correlation method is based on the chemical analysis of the excretory droplets produced during recordings, which allows the origin of the food ingested (phloem or xylem tissue) to be determined. In this method, the correlation could be done (1) by determining the pH of the excreta, where phloem sap has, in general, a higher pH value (7.2–8.5) compared to the xylem sap (5.4–6.5) or (2) by concentration of amino acids. The excretory droplet is applied to a filter paper containing an indicator substance (e.g., ninhydrin or bromophenol blue) to reveal the presence or absence of amino acids. Phloem sap has a higher concentration of amino acids compared to xylem sap. Thus, a color reaction with the indicator compound suggests that the sap is from phloem tissue, whereas no color reaction indicates sap from xylem (Walker 2000 and references therein). For true bugs, this method has not been reported. It is commonly used in auchenorrynchan and sternorrhynchan insects that often excrete droplets.

The correlation process involves multiple methods that often require strenuous effort and skill as well as highly specialized methodology, making the process slow and arduous. In addition, some true bugs, like mirids, do not secrete a salivary sheath, which limits the waveform correlation via histological studies. Furthermore, some plant parts, like the leaf sheaths of grasses, are not amenable to histological processing (Backus et al. 2013). Alternatively, or at least as a complementary method, the biological meanings of the waveforms may be hypothesized based on electrical characteristics, primarily the electrical origins (resistance and electromotive force) of the waveforms by constructing a waveform library. This indirect correlation was used to hypothesize the stylet activities and biological processes of the waveforms for chinch bugs (which are salivary-sheath feeders) (Backus et al. 2013) and for *Lygus* species (cell rupture feeders) (Cervantes et al. 2016).

6.8 Concluding Remarks

Herein we discussed EPG procedures for true bugs, highlighting some problems faced during recordings that needed to be overcome. Perhaps the greatest challenge found was reliably wiring true bugs with larger bodies and restless behavior, such as pentatomids, which culminated in the sanding-and-wiring technique. Moreover, the use of EPG requires some cautions, mostly regarding the effects of electrical settings on feeding behavior, as well as determination and standardization of some pre-recording variables (e.g., access time on host, experimental design, laboratory environmental conditions, etc.).

EPG itself does not directly provide biological meanings of waveforms or specify the plant tissues exploited by insects during each waveform. Indeed, EPG is initially dependent on supplementary methods such as histological studies, artificial diet, and video recording for such correlations. However, once waveforms are identified and these rather time-consuming correlations are completed, the tremendous usefulness of EPG in many applications becomes evident. A deeper, quantitative analysis of EPG waveforms, using statistical comparisons of count and duration data from different feeding activities, can provide valuable information on insect responses to insecticides, resistant crop varieties, transgenic plant toxins, and parasites. These potential uses of the EPG technique will be addressed in Chap. 7.

References

Almeida ACS, Jesus FG, Barrigossi JAF (2020) Unveiling the feeding behavior of *Tibraca limbativentris* (Hemiptera: Pentatomidae) on rice using an electropenetrography waveform library. J Insect Sci 20:14;1–8

Backus EA, Bennett WH (1992) New AC electronic insect feeding monitor for fine-structure analysis of waveforms. Ann Entomol Soc Am 85:437–444

Backus EA, Bennett WH (2009) The AC-DC correlation monitor: new EPG design with flexible input resistors to detect both R and emf components for any piercing-sucking hemipteran. J Insect Physiol 55:869–884

Backus EA, Cline AR, Ellerseick MR, Serrano MS (2007) *Lygus hesperus* (Hemiptera: Miridae) feeding on cotton: new methods and parameters for analysis of nonsequential electrical penetration graph data. Ann Entomol Soc Am 100:296–310

Backus EA, Cervantes FA, Godfrey L, Akbar W, Clark TL, Rojas MG (2018) Certain applied electrical signals during EPG cause negative effects on stylet probing behaviors by adult *Lygus lineolaris* (Hemiptera: Miridae). J Insect Physiol 105:64–75

Backus EA, Rangasamy M, Stamm M, McAuslane HJ, Cherry R (2013) Waveform library for chinch bugs (Hemiptera: Heteroptera: Blissidae): Characterization of electrical penetration graph waveforms at multiple input impedances. Ann Entomol Soc Am 106:524–539

Bonani JP, Fereres A, Garzo E, Miranda MP, Appezzato-da-Gloria B, Lopes JRS (2010) Characterization of electrical penetration graphs of the Asian citrus psyllid, *Diaphorina citri*, in sweet orange seedlings. Entomol Exp Appl 134:35–49

Cervantes FA, Backus EA (2018) EPG waveform library for *Graphocephala atropunctata* (Hemiptera: Cicadellidae): effect of adhesive, input resistor, and voltage levels on waveform appearance and stylet probing behaviors. J Insect Physiol 109:21–40

Cervantes FA, Backus EA, Godfrey L, Akbar W, Clark TL (2016) Characterization of an EPG waveform library for pre-reproductive adult *Lygus lineolaris* and *L. hesperus* feeding on cotton squares. Ann Entomol Soc Am 109:684–697

Cervantes FA, Backus EA, Godfrey L, Rojas MG, Akbar W, Clark TL (2019) Quantitative differences in feeding behavior of *Lygus lineolaris* (Hemiptera: Miridae) on transgenic and nontransgenic cotton. J Econ Entomol 112:1920–1925

Chinchilla-Ramírez M, Garzo E, Fereres A, Gavara-Vidal J, ten Broeke CJM, van Loon JJA, Urbaneja A, Pérez-Hedo M (2021) Plant feeding by *Nesidiocoris tenuis*: Quantifying its behavioral and mechanical components. Biol Control 152:104402;1–10

Cline AR, Backus EA (2002) Correlations among AC electronic monitoring waveforms, body postures, and stylet penetration behaviors of *Lygus hesperus* (Hemiptera: Miridae). Environ Entomol 31:538–549

Jin S, Chen ZM, Backus EA, Sun XL, Xiao B (2012) Characterization of EPG waveforms for the tea green leafhopper, *Empoasca vitis* Göthe (Hemiptera: Cicadellidae), on tea plants and their correlation with stylet activities. J Insect Physiol 58:12351244

Joost PH, Backus EA, Morgan D, Yan F (2006) Correlation of stylet activities by the glassy-winged sharpshooter, *Homalodisca coagulata* (Say), with electrical penetration graph (EPG) waveforms. J Insect Physiol 52:327–337

Kindt F, Joosten NN, Peters D, Tjallingii WF (2003) Characterization of the feeding behaviour of western flower thrips in terms of electrical penetration graph (EPG) waveforms. J Insect Physiol 49:183–191

Li W, Wyckhuys KAG, Wu K (2016) Does feeding behavior of a zoophytophagous mirid differ between host plant and insect prey items? Arthropod Plant Interact 10:79–86

Lu Z, Dong S, Li C, Li L, Yu Y, Men X, Yin S (2020) Sublethal and transgenerational effects of dinotefuran on biological parameters and behavioural traits of the mirid bug *Apolygus lucorum*. Sci Rep 10:226;1–8

Lucini T, Panizzi AR (2016) Waveform characterization of the soybean stem feeder *Edessa meditabunda*: overcoming the challenge of wiring pentatomids for EPG. Entomol Exp Appl 158:118–132

Lucini T, Panizzi AR (2017a) Feeding behavior of the stink bug *Dichelops melacanthus* Dallas on corn seedlings: an EPG analysis at multiple input impedances and histology correlation. Ann Entomol Soc Am 110:160–171

Lucini T, Panizzi AR (2017b) Probing behavior of *Dichelops furcatus* (F.) (Heteroptera: Pentatomidae) on wheat plants characterized by electropenetrography (EPG) and histological studies. J Insect Sci 17:65;1–15

Lucini T, Panizzi AR (2018a) Electropenetrography (EPG): a breakthrough tool unveiling stink bug (Pentatomidae) feeding on plants. Neotrop Entomol 47:6–18

Lucini T, Panizzi AR (2018b) Electropenetrography monitoring of the neotropical brown-stink bug (Hemiptera: Pentatomidae) on soybean pods: an electrical penetration graph-histology analysis. J Insect Sci 18:5;1–14

Lucini T, Panizzi AR (2020) Electropenetrographic comparison of feeding behavior of *Dichelops furcatus* (Hemiptera: Heteroptera: Pentatomidae) on soybean and spring cereals. J Econ Entomol 113:1796–1803

McLean DL, Kinsey MG (1964) A technique for electronically recording aphid feeding and salivation. Nature 202:1358–1359

Mitchell PL, Cooke SB, Smaniotto LF (2018) Probing behavior of *Nezara viridula* on soybean: characterization and comparison of electrical penetration graph (EPG) waveforms on vegetative and reproductive plant structures. J Agric Urban Entomol 34:19–43

Rangasamy M, McAuslane HJ, Backus EA, Cherry RH (2015) Differential probing behavior of *Blissus insularis* (Hemiptera: Blissidae) on resistant and susceptible St. Augustinegrasses. J Econ Entomol 108:780–788

Rivera KL, Mitchell PL (2020) Probing behavior of *Nezara viridula* first instars: EPG analysis and effect of food availability on subsequent development. J Agric Urban Entomol 36:47–63

Sarria E, Cid M, Garzo E, Fereres A (2009) Workbook for automatic parameter calculation of EPG data. Comput Electron Agric 67:35–42

Serteyn L, Ponnet L, Saive M, Fauconnier ML, Francis F (2020) Changes of feeding behavior and salivary proteome of brown marmorated stink bug when exposed to insect-induced plant defenses. Arthropod Plant Interact 14:101–112

Spiller NJ, Koenders L, Tjallingii WF (1990) Xylem ingestion by aphids – a strategy for maintaining water balance. Entomol Exp Appl 55:101–104

Stubbins FL, Mitchell PL, Turnbull MW, Reay-Jones FPF, Greene JK (2017) Mouthpart morphology and feeding behavior of the invasive kudzu bug, *Megacopta cribraria* (Hemiptera: Plataspidae). Invert Biol 136:309–320

Tjallingii WF (1978) Electronic recording of penetration behaviour by aphids. Entomol Exp Appl 24:721–730

Tjallingii WF (1988) Electrical recording of stylet penetration activities. In: Minks AK, Harrewijn P (eds) World crop pests: aphids, their biology, natural enemies and control. Elsevier, Amsterdam, pp 95–108

Tuelher ES, Backus EA, Cervantes F, Oliveira EE (2020) Quantifying *Lygus lineolaris* stylet probing behavior and associated damage to cotton leaf terminals. J Pest Sci 93:663–677

Van Helden M, Tjallingii WF (2000) Experimental design and analysis in EPG experiments with emphasis on plant resistance research. In: Walker GP, Backus EA (eds) Principles and applications of electronic monitoring and other techniques in the study of homopteran feeding behavior. Lanham, Entomological Society of America, pp 144–171

Walker GP (2000) A beginner's guide to electronic monitoring of homopteran probing behavior. In: Walker GP, Backus EA (eds) Principles and applications of electronic monitoring and other techniques in the study of homopteran feeding behavior. Lanham, Entomological Society of America, pp 14–40

Chapter 7
Role of EPG in Developing and Assessing Control Methods for Heteropteran Crop Pests

Contents

Abstract The multiple interactions of the electrical penetration graph (EPG) technique with various strategies that can be used to manage pests of crop plants are presented and discussed in this chapter. We here include insecticides, still the main tool in use to control true bug pests worldwide by either contact action or systemic circulation in plants; plants that show resistance in the broad sense, either by traditional plant breeding or by advanced genetically modified (GM) techniques; natural enemies, including parasitoids, pathogens, and zoophytophagous predatory Heteroptera; and gene silencing by interference RNA. These strategies, with the potential to control heteropterans and other crop pests, will benefit from the information that the EPG technique can provide, enhancing their efficiency as pest control measures.

Keywords Feeding sites · Insecticides · Host plant resistance · Toxins · Transgenic plants · RNAi

A. R. Panizzi et al., *Electronic Monitoring of Feeding Behavior of Phytophagous True Bugs (Heteroptera)*, Entomology in Focus 6,
https://doi.org/10.1007/978-3-030-64674-5_7

7.1 Introduction

Heteropterans have long been recognized as major pests of cultivated plants all over the world. Their importance in causing damage to crops is legendary, and this has been emphasized in publications worldwide. However, it was not until the year 2000 that a book dedicated exclusively to true bugs was published, highlighting their global importance as pests. This book covers the main pest species of cultivated plants and those of medical importance, as well as those that have potential as biological control agents (Schaefer and Panizzi 2000). Other books were published covering specific taxa and geographic areas (e.g., Stink Bugs [Pentatomidae] of Economic Importance in America North of Mexico [McPherson and McPherson 2000]).

True bugs feed by inserting their stylets (mandibles + maxillae) into plant tissues to obtain water and nutrients. The outer mandibles with their teeth-like structures tear the tissues to allow the stylets to penetrate, often causing substantial mechanical (physical) damage, especially by true bugs using the cell-rupture feeding strategy. This may be followed by chemical damage due to the action of injected enzymes. Finally, as plants react to the physical and chemical disruption, hormonal and physiological imbalances occur, leading to malformation of fruits and vegetative tissues and abscission of reproductive structures (Hori 2000 – see illustration in Fig. 7.1).

True bugs explore many different feeding sites on the host plant, which will be briefly discussed ahead (this can be seen in greater detail in Chap. 3). Control methods have been investigated to mitigate the impact of these important pests on

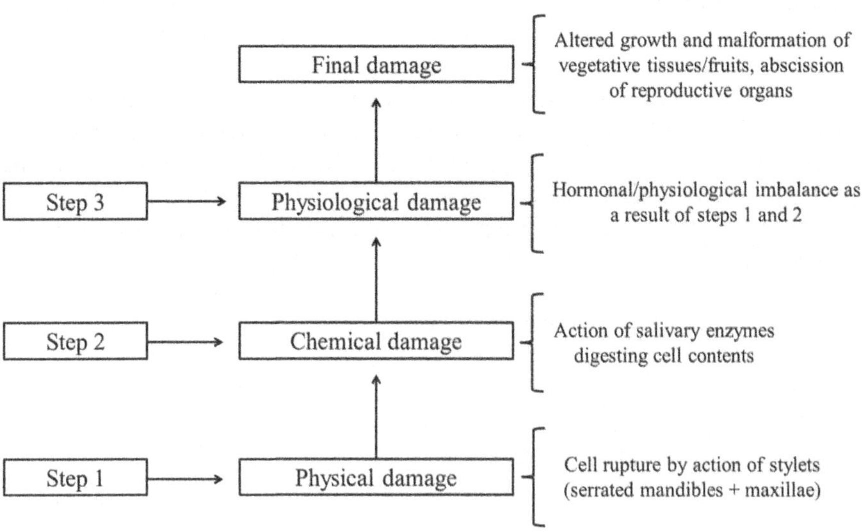

Fig. 7.1 Main steps during the cell-rupturing feeding process of true bugs on plants that causes several types of damage and results in final damage as a summation of the previous steps. (Based on Hori [2000])

cultivated plants. These include mostly insecticides with different modes of action. Plant resistance has also been an important strategy to manage pest species, particularly those plants bearing toxins expressed via transgenic methods. Rccently, the technology of gene silencing via RNAi is gaining momentum as a modern tool to manage heteropteran pests. These control methods will be discussed, with a focus on electrical penetration graph (EPG) technique as a tool to evaluate their efficiency and improve their efficacy to manage pest species.

7.2 Feeding Sites

Phytophagous heteropterans feed by inserting their stylets into the plant issues to suck out cell contents. Feeding sites include vascular tissues (xylem and phloem), mesophyll/parenchyma, inflorescences/buds, and seed reserve tissue (Fig. 7.2). Ingestion from xylem is mostly to maintain body hydration (Spiller et al. 1990; Rivera and Mitchell 2020) and for balance of nutrient concentration purposes (Lucini et al. 2016), since the sap is highly water concentrated and diluted in nutrients (Taiz and Zeiger 2004). EPG studies, mostly coupled with histology, have demonstrated ingestion from xylem vessels for several species of true bugs (blissids, coreids, and pentatomids). In particular, many species of pentatomids that feed on seed endosperm (highly concentrated in nutrients and the most preferred food source) have been shown to hydrate from xylem (e.g., *Piezodorus guildinii* (Westwood) [Lucini et al. 2016]; *Diceraeus (Dichelops) furcatus* (F.) [Lucini and

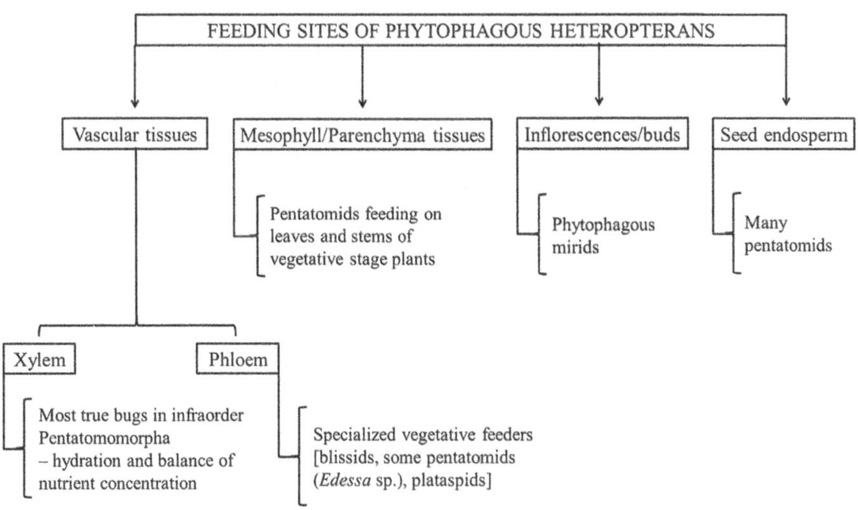

Fig. 7.2 Summary of the ingestion sites exploited by phytophagous heteropterans determined using the EPG technique

Panizzi 2017b]; *Euschistus heros* (F.) [Lucini and Panizzi 2018]; and *Nezara viridula* (L.) [Mitchell et al. 2018]).

For other species of true bugs, specialized to feed on vegetative structures, stylet tips have been found or proposed to be inserted into phloem tissue, such as the pentatomid *Edessa meditabunda* (F.) (Lucini and Panizzi 2016), the chinch bugs *Blissus insularis* Barber and *Blissus occiduus* Barber (Backus et al. 2013), and the plataspid *Megacopta cribraria* (F.) (Stubbins et al. 2017). The cells of mesophyll and stem parenchyma tissues have also been exploited as a feeding site, as demonstrated for the stink bugs, *Diceraeus (Dichelops) melacanthus* (Dallas) (Lucini and Panizzi 2017a), *Halyomorpha halys* Stål (Serteyn et al. 2020a), and *Tibraca limbativentris* Stål (Almeida et al. 2020). EPG also showed that the mirid *Lygus lineolaris* (Palisot de Beauvois) fed on mesophyll/parenchyma tissues of cotton leaves (Tuelher et al. 2020); in addition, *Lygus* spp. are closely associated with feeding on flower buds of cotton plants (e.g., Cervantes et al. 2016, 2017).

Why is it important to know the feeding sites of phytophagous heteropterans? As management strategies are developed, including systemic insecticides circulating in the plant tissues or toxins expressed by transgenic plants, this information will be crucial. As we learn about the frequency and duration of feeding events at various feeding locations, expression of the toxic materials at these sites could become a selective and efficient control strategy. The use of EPG, generating feeding waveforms, coupled with histology creates a complete set of information that will tell us how bugs will be exposed to toxins and how they will respond. This response includes how the feeding process is altered and at what point bugs will cease their ingestion activities which will eventually lead to their death.

7.3 Insecticides

Insecticide activity has long been known to involve multiple actions and reactions of the toxicant and the insect tissues, including penetration of the toxicant and activation of the target site, followed by detoxification (Sun 1968). The post-treatment period includes the reactions of insects in several stages in a sequence, as follows: normal activity, hyperactivity, lack of coordination, prostration and convulsions, general paralysis, and apparent death (Roan and Hopkins 1961). Several aspects of how insecticides intoxicate insects leading to their death have been presented and discussed over the years (e.g., O'Brien 1966; Bloomquist 1996; Casida and Quistad 1998; Zlotkin 1999; Casida and Durkin 2013; Casida 2018), but for the more restrictive purpose of this chapter, we will consider just the two main modes of exposure: contact and systemic. Our scope is to evaluate the insect responses to insecticides using the EPG technique and to elucidate how different exposures to insecticides will affect their feeding behavior.

7.3.1 Contact Insecticides

Insecticides that fall in this class are those that are absorbed through the external body surface of the insect. They hit the target through spray, or bugs walk on a surface that has received the insecticide, and the chemicals are absorbed through the tarsi. Information on their mode of entry has been available for many years (e.g., Gerolt 1969), and the effect of contact insecticides on true bugs has been studied worldwide since their discovery in the late 1940s. However, despite the myriad studies on true bugs vs. contact mode insecticides, to our knowledge, the use of EPG to investigate this interaction has never been attempted. This is in part because monitoring and recording of true bug feeding is a relatively recent application of the EPG technique. The pioneering studies were focused on defining the behavioral activities of waveforms and correlating them with the different feeding sites (e.g., Cline and Backus 2002; Backus et al. 2013; Lucini et al. 2016). As studies with true bugs and EPG progressed, new ideas and research goals started to emerge, and one of these is to monitor and record the effects of contact insecticides on feeding behavior.

For example, studies have shown that the use of sodium chloride (NaCl) enhances the effects of contact insecticides (Corso and Gazzoni 1998). This occurs due to the increased time that bugs stay on the plant surface treated with NaCl solution, thereby increasing their food contact behavior (Niva and Panizzi 1996), and then their exposure to the insecticide, which increases mortality. This increase in the food contact was recorded by visual examination, which was time consuming and hard to achieve. Now, consider this research done using the EPG technique; the information could be achieved in an easier and more effective way, with less probability of error.

7.3.2 Systemic Insecticides

Systemic insecticides are applied either through spray on the plant foliage, placement in the soil to be absorbed through the roots, or through seed treatment before planting. Once in contact with the plants, they are absorbed and circulate in the plant. Sucking insects are poisoned when they pierce the plant tissues and ingest cell contents. Among the systemic insecticides, the recently discovered and commercialized neonicotinoids (neonics) are highly efficient at controlling piercing-sucking insects; the most common ones are imidacloprid and thiamethoxam. This group of insecticides, although in general readily metabolized, has been highly criticized because of their toxic effect on honey bees and other pollinators (Casida 2018).

Few detailed studies are available to fully elucidate how true bugs react to alter their feeding activities in the presence of systemic insecticides in different feeding sites. This technology has been largely applied on other piercing-sucking hemipterans to study the disruption caused by systemic insecticides on their feeding behavior (e.g., Serikawa et al. 2012; Jacobson and Kennedy 2014; Mustafa et al. 2015; Tariq et al. 2017; Waqas et al. 2019). Only recently, a study with the mirid bug *Apolygus lucorum* (Meyer-Dür) was conducted to evaluate the effect of a neonicotinoid on its behavioral traits using EPG (Lu et al. 2020). At this point, the EPG clearly shows its value; it can monitor and record, via waveform analyses, how the number and duration of feeding events are affected by insecticide exposure.

Moreover, once the feeding sites are identified and the feeding sessions in each one fully characterized, systemic insecticides could, in theory, be directed to circulate in the sites where bugs spend more time ingesting (mostly vascular tissues). This information might be of use to increase the efficacy of insecticides, allowing use of lower dosages and mitigating the appearance of resistant strains of the bugs under control.

7.4 Predators, Parasitoids, Entomopathogens, and Endosymbionts

Zoophytophagous mirids have potential as biological control agents, but understanding their feeding behavior on crop plants, and the consequent damage, is essential (Sánchez and Lacasa 2008). EPG offers an ideal method to determine not only the manner in which these omnivorous insects attack plant tissue but also the extent of their feeding. Two species of zoophytophagous mirids considered as biocontrol agents have been examined using EPG. *Nesidiocoris tenuis* (Reuter) use the cell rupturing strategy, ingesting from parenchyma and vascular tissues on tomato plants; fifth instars had greater damage potential than adults (Chinchilla-Ramírez et al. 2021). *Apolygus lucorum* (Meyer-Dür) was recorded using EPG on green beans and lepidopteran eggs (Li et al. 2016). When plant and insect foods were presented together, overall feeding was considerably longer on bean pods than eggs. A similar pattern was found when foods were presented separately: both overall feeding and single feeding events were longer on green bean (Li et al. 2016). This type of information is vital in determining the potential role of omnivorous mirids as predators and the trade-off between beneficial and damaging behaviors in specific agroecosystems.

As is widely known, phytophagous true bugs are susceptible to parasitoids that attack both eggs and nymphs/adults. In the latter case, the impact of the natural enemies on the performance of these heteropteran crop pests must be assessed. For example, the tachinid *Trichopoda giacomelli* (Blanchard) (= *Eutrichopodopsis nitens* Blanchard) is a common parasite of the southern green stink bug, *N. viridula*, in South America. When it parasitizes late instar nymphs or early adults, a drastic reduction in longevity is observed, and females do not reproduce; when parasitism occurs on older adults (1-week-old), reproduction is reduced by about 60% (Corrêa-Ferreira et al. 1991), and adults die after larval emergence. Tachinid flies are, in general, much more abundant as true bug parasites (mostly on pentatomids) than reported, mainly because they often lay their eggs underneath the bug's wings where they are overlooked during field observations (e.g., Eger 1981; Aldrich et al. 2006; Agostinetto et al. 2018; Lucini et al. 2020).

Other adult parasites of true bugs include hymenopterans (e.g., *Hexacladia*) that affect the bug's performance (i.e., longevity, reproduction) and kill them upon emergence (e.g., Corrêa-Ferreira et al. 1998; Panizzi and Silva 2010). In this case it is difficult to know if the bug is parasitized or not, because the wasp lays eggs internally in the bug's body. To measure their effect on bug biology, laboratory tests must be conducted by artificially parasitizing the bugs and comparing their performance with healthy ones.

Entomopathogenic agents (fungi, bacteria, viruses, and nematodes) have been extensively studied as biological control agents (e.g., Tanada 1959; Moscardi 1999; Liu et al. 2000; Baverstock et al. 2010; Karabörklü et al. 2018). Research has dealt with how the target insect detects and avoids the pathogen and how the transmission of pathogens between host insects occurs. The behavioral interactions between insects and entomopathogenic agents, from commercially available biopesticides to non-formulated, naturally occurring pathogens, have been widely investigated (e.g., Baverstock et al. 2010). Multiple isolates of entomopathogenic fungi have shown potential to be used for microbial control against heteropterans, such as the mirid *L. lineolaris* (Liu et al. 2002), the plataspid *M. cribraria* (Seiter et al. 2014, Britt et al. 2016), and several pentatomid species, including *H. halys* (Gouli et al. 2012), *Bagrada hilaris* Burmeister (Barrera-López et al. 2020), *P. guildinii* (Parys and Portilla 2020) *E. heros*, and *D. furcatus* (Resquín-Romero et al. 2020).

Mermithid nematodes (e.g., *Agamermis* spp., *Hexamermis* spp., and *Mermis* spp.) have been also reported parasitizing heteropterans, for example, *M. cribraria* (Stubbins et al. 2015, 2016), and the pentatomids *Chinavia hilaris* (Say) (Kamminga et al. 2012), and *P. guildinii* (Ribeiro and Castiglioni 2009, Kamminga et al. 2012). Furthermore, bacterial strains have showed potential for management of *N. viridula* (Martin et al. 2007) and *H. halys* (Tozlu et al. 2019), as well as viruses, which were reported on *N. viridula* (Williamson and von Wechmar 1995).

Other interactions with microorganisms include endosymbionts and how they affect their hosts. Their presence in certain cases are obligate for both partners, and their elimination cause insects to grow poorly and produce few or no offspring (e.g., Douglas 1998). In true bugs, several studies have highlighted the important role of endosymbionts to their biology in general. Recently, with advances in molecular techniques, these associations have become clearer, improving our understanding of the bug/symbiont associations (Prado and Zucchi 2012, and references therein). In certain cases, however, elimination of microorganisms living in the gut using antibiotics can accelerate nymphal development and increase adult survivorship and longevity, as demonstrated for the stink bug *N. viridula* (Hirose et al. 2006). In this last instance, apparently, microorganisms that were detrimental to the bug biology were eliminated. Clearly, all these interactions need further, detailed study to be fully understood.

These few examples illustrate the impact of natural enemies and endosymbionts on true bug biology. In general, studies with natural enemies aim to determine whether they kill the host, how quickly the host dies, and their overall impact on the pest population. Endosymbiont research has concentrated on identification and how their elimination affects nymphal development and adult reproduction and longevity. Few studies have focused on the effects of natural enemies or endosymbionts on bug behavior, for example, how feeding frequency and duration of feeding sessions are affected. One may wonder, why so? We believe that this may be attributed mostly to the lack of an efficient and practical technology that facilitates this type of study. In any case, we think that to fill this gap the use of EPG is the answer.

7.5 Plant Resistance to True Bugs

Plant resistance to insects has been investigated since the pioneering works published over 70 years ago by Snelling (1941) and Painter (1951) that set the foundations for what later became known as host plant resistance. Plants are considered to have evolved tolerance (to suffer less negative impact from herbivore damage) and resistance (to prevent or reduce damage through physical defenses and production of secondary metabolites) as defense strategies against herbivores (see Núñez-Farfan et al. 2007; Orians and Ward 2010; Kariñho-Betancourt and Núñez-Farfán 2015, and references therein).

Cultivated plants, in particular, have obtained resistance in the broad sense by means of experimental plant breeding. Painter, considered the father of host plant resistance, described in 1951 three main mechanisms of resistance – antibiosis, tolerance, and non-preference (antixenosis) (but see Stout [2013] on the revision of this trichotomous framework, reducing it to a dichotomous scheme: resistance [plant traits that limit injury to the plant subdivided into constitutive/inducible and direct/indirect subcategories] and tolerance [plant traits that reduce amount of yield loss per unit injury]). A myriad of research efforts on different cultivated plants and their associated insect pests followed, to be published in journals and compiled in text books worldwide (e.g., Panda 1979; Maxwell and Jennings 1980; Smith 1989; Dhaliwal and Singh 2004; Baldin et al. 2019, just to mention a few).

As a case study, let us examine the evolution of a major commodity, soybean, and its resistance to stink bugs that evolved over time as a result of joint efforts by plant breeders and entomologists. These bugs form a complex of pest species that damage the crop worldwide and in particular in the Americas, the world greatest soybean production area (e.g., Panizzi and Slansky 1985; Panizzi et al. 2000; Smaniotto and Panizzi 2015).

Different sources of soybean genotypes with multiple resistance to insects were discovered in the 1970s in the USA (Duyn et al. 1971; Clark et al. 1972; Luedders and Dickerson 1977). As a result, several research programs were launched testing the lines 'PI 171451,' 'PI 227687,' and 'PI 229358' and genetic materials derived from them, for resistance to stink bugs (e.g., Jones and Sullivan 1979; Lourenção et al. 1987; Rossetto et al. 1995; McPherson et al. 2007; Campos et al. 2010; Silva et al. 2013, 2014). These programs and others yielded several materials (e.g., cv. 'IAC 100') that later were released as commercial cultivars. However, they did not succeed commercially because of lack of suitable agronomic characteristics and/or low productivity (seed yield).

More recently, a new soybean cultivar 'BRS 1003IPRO' was released by the Embrapa Soybean Research Center, at Londrina, Paraná, Brazil. The cultivar is said to tolerate over twice the existing stink bug economic injury level (2 bugs/m) with no significant reduction in seed yield, compared with standard cultivars (Carlos Arrabal Arias, personal communication to ARP). However, no published data are available on the mechanisms of resistance/tolerance to the stink bugs, or whether antibiosis or antixenotic effects are operating. Clearly, further investigations are needed to fully elucidate the mechanisms of resistance of the 'BRS 1003IPRO' cultivar and other newly developed resistant lines. Electronically monitoring stink bug responses to these cultivars in comparison to standard (conventional) ones would be the ideal method of investigation, as shown by previous EPG studies in other crops and with different hemipteran pests.

The EPG technique has been extensively applied to study the behavioral characteristics of several species of piercing-sucking insects on resistant plants/genotypes (e.g., Alvarez et al. 2006; Diaz-Montano et al. 2007; Ghaffar et al. 2011; Miao et al. 2014; Todd et al. 2016; Koch et al. 2018; Sun et al. 2018a). To our knowledge, only two works have been published on true bugs to evaluate plant resistance using EPG. The first one deals with comparing resistant versus susceptible St. Augustine grasses to the chinch bug *B. insularis* (Hemiptera: Blissidae) (Rangasamy et al. 2015). These authors found that bugs spent less time feeding on resistant 'FX-10' and 'NUF-76' than on susceptible 'Floratam' and 'Palmetto' cultivars. Moreover, the feeding events were of shorter duration, although more numerous, in resistant compared with susceptible cultivars; pathway-associated stylet activities (i.e., penetration through epidermal and mesophyll tissue) took longer in the resistant varieties than in the susceptible cultivars; and bugs spent less time in xylem and phloem ingestion in resistant than in susceptible cultivars.

The second work was recently conducted with the invasive brown marmorated stink bug, *H. halys* (Serteyn et al. 2020b). These authors found that lipoxygenase activity was enhanced in faba bean, *Vicia faba* L. leaves by previous bug feeding. The EPG analysis of *H. halys* feeding behavior on elicited plants showed that initial and sustained ingestion probes were delayed and ingestion events were shorter. Furthermore, EPG recordings of feeding behavior could be directly associated with changes in the bugs' salivary gland proteins.

The tremendous potential of the EPG technique to investigate host plant resistance to true bugs has not yet been properly explored. The use of EPG to monitor heteropteran feeding has thus far primarily emphasized qualitative studies to define the biological meanings of waveforms, which must be completed before quantitative comparisons can be made. Therefore, as such qualitative studies are now being completed for an increasing number of true bug species, we believe that EPG will become the main tool to be used in host plant resistance studies in the near future.

7.6 Toxins Expressed by Transgenic Plants

Genetically modified (GM) plants with resistance to arthropod pests started to become popular during the last 20–30 years. Several molecular genetic tools and approaches were developed, creating a wide diversity of resistant germplasm and crop cultivars (e.g., Smith et al. 1994; Yencho et al. 2000; Sadasivam and Thayumanavan 2003; Smith and Clement 2012).

Perhaps the most common traits explored are the transgenes from *Bacillus thuringiensis* Berliner (*Bt*), which encode for insecticidal crystalline proteins (Cry). They apparently act by creating pores that disrupt epithelial membranes of the gut of juvenile insects and allow pathogens entrance into the hemocoel causing the death of the insect. GM varieties bearing Cry proteins are widely used in major cropping systems and have been subjected to regulation and assessed for human and environmental safety (Latham et al. 2017).

Transgenic *Bt*-crops such as cotton, maize, soybean, and rice cultivars expressing resistance to several lepidopteran pests became major components of agriculture (Qaim and Zilberman 2003; Wang et al. 2016; Marques et al. 2017; Chen et al. 2017; Eghrari et al. 2019; Chakrabarty et al. 2020). EPG has been used to evaluate the effect of *Bt*-crops on many piercing-sucking insects, including aphids, whiteflies, and planthoppers (e.g., Liu et al. 2005; Xue et al. 2009; Yin et al. 2010; Sun et al. 2018b). Regarding true bugs, only two studies used EPG to evaluate insect-resistant cultivars bearing toxins (Abbate et al. 2019; Cervantes et al. 2019).

However, the impact of *Bt*-crops on some true bugs seems to be minimal. At least one commercially available *Bt* soybean, which was tested against the red-banded stink bug, *P. guildinii*, showed no significant detrimental effects on feeding behavior (Abbate et al. 2019). These authors used EPG to compare *Bt* versus non-*Bt* cultivars and found differences only in the waveform related to pathway (stylet penetration into plant tissue), which was shorter in the *Bt* cultivar. No differences were observed in waveforms related to feeding sites (xylem and seed endosperm). These differences in stylet pathway duration are believed to be associated with morphological differences between the cultivars tested, rather than due to the presence of the Cry endoprotein (Abbate et al. 2019).

Bt-expressing cotton cultivars have been reported to affect the feeding behavior of nymphs of the mirid bug *L. lineolaris* (Cervantes et al. 2019). These authors found that ingestion events were shorter on transgenic plants. Results suggest that *Bt* cultivars are less palatable and/or preorally digestible and that the presence of Cry protein could interfere with complete dissolution (liquefaction) of the cotton tissues via enzymatic activity (cell rupturing strategy) for subsequent ingestion. These contrasting results on the effect of *Bt* on piercing-sucking phytophagous insects clearly suggest additional studies are needed, and EPG will continue to be the best tool choice.

Besides *Bt*, other genes for proteins exhibiting toxicity to insects, such as proteinase inhibitors, α-amylase inhibitors, lectins, and chitinases, have been expressed in the genomes of many crop plants (Smith and Clement 2012). Moreover, abiotic factors such as sunlight may enhance the effects of allelochemicals in plants. For example, solar UV-B radiation apparently potentializes isoflavonoid content of soybean pods, increasing resistance of plants to stink bugs (Zavala et al. 2015).

These and other toxins in cultivated plants have been evaluated against most feeding guilds of insect pests using standard (classic) biological studies in the laboratory and in the field. However, many of these toxins have proven ineffective against true bugs; apparently, they are either less or non-toxic or the bugs are selective and able to avoid them during the feeding process. In any case, studies with EPG should generate new and relevant information regarding these poorly known interactions.

7.7 Gene Silencing by RNAi and Implications for True Bug Control

RNA interference (RNAi), or post-transcriptional gene silencing, is a promising new technique for control of insect pests (Price and Gatehouse 2008; Huvenne and Smagghe 2010), in which selected genes essential for cell functioning are downregulated by severing messenger RNA (mRNA), thereby preventing translation. The process within a cell involves double-stranded RNA (dsRNA), which is cleaved into shorter interference RNAs (siRNA), which in turn become part of an RNA interference silencing complex which binds to, and cuts, the target mRNA (Huvenne and Smagghe 2010; Joga et al. 2016). Once the molecular mechanisms of RNA interference were described (see Wilson and Doudna 2013), the potential applications of RNAi for pest management began to gain momentum (see reviews by Cooper et al. 2019; Dias et al. 2020; Zhu and Palli 2020; Christiaens et al. 2020).

Only recently has research been done to investigate application of this technique to Heteroptera. Thus far, seven species have been found to possess the requisite cellular core RNAi machinery and/or respond to exogenously applied dsRNA: the southern green stink bug, *N. viridula* (Davis-Vogel et al. 2018; Gurusamy et al. 2020; Riga et al. 2020; Sharma et al. 2021); the brown marmorated stink bug, *H. halys* (Mogilicherla et al. 2018); the Neotropical brown stink bug, *E. heros* (Castellanos et al. 2019; Cagliari et al. 2020); the green-belly stink bug, *D. melacanthus* (Pinheiro et al. 2020); the harlequin bug, *Murgantia histrionica* (Hahn) (Howell et al. 2020); the tarnished plant bug, *L. lineolaris* (Walker and Allen 2010, 2011); and the hematophagous bed bug, *Cimex lectularius* L. (Basnet and Kamble 2018a, b).

Microinjection was used to apply dsRNA during initial studies of the RNAi mechanism in insects; for pest management this is impractical (Joga et al. 2016). Instead, the target insect ingests the dsRNA, by consuming free liquid or feeding on plants. The salivary environment in the pre-oral cavity must not be damaging to the dsRNA, which needs to reach the midgut intact and pass from the lumen through the midgut lining. Uptake of cells in the gut epithelium is only the first step; the dsRNA ideally should spread to other tissues, thereby allowing cells other than those of the midgut to be targeted to create a systemic silencing (Huvenne and Smagghe 2010). Thus, intracellular RNAi machinery, cellular uptake, transport mechanisms, alimentary canal environment, and delivery methods may all affect the success of the RNAi process when used for management of a pest species. Coleoptera tend to be highly responsive to ingested dsRNA (Christiaens et al. 2020), whereas Lepidoptera and Diptera show more variability (Cooper et al. 2019).

Methods to introduce dsRNA other than injection include consumption of treated droplets, root uptake, trunk injection, soaking of food material, and transgenic plants; the latter was first demonstrated for corn rootworm on maize (Baum et al. 2007) and now holds considerable commercial promise. Droplets of treated liquid are most readily consumed by lepidopteran larvae, but this method, using dsRNA in sucrose solution, was effective for *M. histrionica* (Howell et al. 2020). In general, for heteropterans, root uptake and immersion methods have been the most investigated. A protocol for delivery of dsRNA by soaking green beans in liquid has recently been developed for brown marmorated stink bug (Ghosh et al. 2017); the bean vascular system effectively delivered functional dsRNA and expression of targeted genes was reduced. Further studies using this delivery system with other target genes successfully produced high mortality in *H. halys* second instars (Mogilicherla et al. 2018). Root uptake was investigated using mustard seedlings and harlequin bugs; roots were immersed in water containing dsRNA and *M. histrionica* second instars were allowed to feed for 3 days. High mortality was achieved with several target genes, and the dsRNA was stable in bug saliva and within the seedlings (Howell et al. 2020).

The various oral delivery systems being developed – transgenic crops, immersion, and soil uptake – depend on active insect feeding for effective plant-mediated delivery of an optimal quantity of dsRNA. The amount of dsRNA encountered by the insect is important, as silencing is concentration dependent (Baum et al. 2007) but exceeding the optimal concentration does not increase effectiveness (Joga et al. 2016). Therefore, the potential for EPG in assessing feeding activity in the development of plant-mediated delivery cannot be overstated. Concentration coupled with duration of feeding will determine dsRNA intake; EPG, unlike visual observations, can distinguish between apparent feeding (stylets inserted in tissue) and the period of actual ingestion.

In most studies of RNAi efficacy against insects, the evaluation method for gene expression is mortality, but in a few cases, depending on the target gene silenced, food uptake may be measured. Here, again, EPG will be of use in assessing changes in feeding behavior and marking the cessation of feeding. Pathway waveforms can reveal the number of feeding attempts, reductions in frequency or duration of ingestion can be measured, and if genes involved with watery or gelling saliva production can be silenced, EPG provides a method of indirectly visualizing the salivation process. Thus, for monitoring intake of dsRNA as plant-mediated delivery methods are developed, and for evaluating insect responses to gene silencing, EPG has an important role to play in this new emerging technology.

7.8 Concluding Remarks

In this chapter we have presented and discussed the actual and potential interactions of the EPG technique with tactics that can be used to manage true bug pests of crop plants. These strategies include the traditional contact and systemic insecticides, still widely used to control heteropteran pests; plant resistance either via classical plant breeding or via genetically modified (GM) plants; natural enemies; and the latest tool of gene silencing by RNA interference. Figure 7.3 summarizes and illustrates some of these interactions. In general, the impact of these strategies on true bugs requires more detailed elucidation, which the electrical penetration graph technique may provide.

The sophisticated feeding habits of true bugs and their resulting damage to crop plants challenge our understanding of this interaction. The physical (mechanical) damage imposed on plants and the chemical damage caused by salivary enzyme activity of heteropterans (see Fig. 7.1) have been investigated by numerous authors worldwide. Basic studies using tools such as EPG are essential to understand fully the multiple interactions that result in the final damage, as recently illustrated by the use of EPG to quantify the feeding behavior of *L. lineolaris* nymphs and adults on cotton leaflets and then correlate it with the associated damage (Tuelher et al. 2020). In conclusion, it is clear that the role of EPG in developing and assessing control methods for heteropteran crop pests has a bright future.

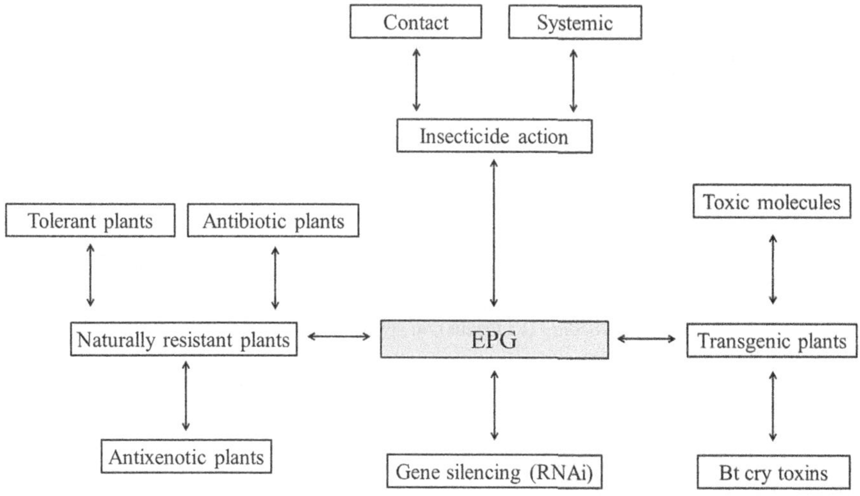

Fig. 7.3 Possible multiple interactions of EPG with natural and transgenic (genetically modified – GM) plants, the gene silencing technique by RNA interference (RNAi), and insecticide actions to monitor the feeding of phytophagous insect pests and other behavioral responses

References

Abbate S, Lucini T, Bonansea S, Bentancur O, Altier N, Pons X, Panizzi AR (2019) Bt soybean does not affect feeding behavior of red-banded stink bug *Piezodorus guildinii* (Hemiptera: Pentatomidae). XIX international plant protection congress, Hyderabad, India, Abstract

Agostinetto A, Panizzi AR, Lucini T (2018) Monthly parasitism rate by tachinid flies and egg allocation on the body of *Dichelops furcatus*. Fla Entomol 101:91–96

Aldrich JM, Khrimian A, Zhang A, Shearer PW (2006) Bug pheromones (Hemiptera, Heteroptera) and tachinid fly host-finding. Denisia 50:1015–1031

Almeida ACS, Jesus FG, Barrigossi JAF (2020) Unveiling the feeding behavior of *Tibraca limbativentris* (Hemiptera: Pentatomidae) on rice using an electropenetrography waveform library. J Insect Sci 20:14;1–8

Alvarez AE, Tjallingii WF, Garzo E, Vleeshouwers V, Dicke M, Vosman B (2006) Location of resistance factors in the leaves of potato and wild tuber-bearing *Solanum* species to the aphid *Myzus persicae*. Entomol Exp Appl 121:145–157

Backus EA, Rangasamy M, Stamm M, McAuslane HJ, Cherry R (2013) Waveform library for chinch bugs (Hemiptera: Heteroptera: Blissidae): characterization of electrical penetration graph waveforms at multiple input impedances. Ann Entomol Soc Am 106:524–539

Baldin, ELL, Vendramim JD, Lourenção AL (2019) Plant resistance to insects. Fundamentals and applications. University of São Paulo Press, FEALQ, Piracicaba, São Paulo (*in Portuguese*)

Barrera-López AA, Guzmán-Franco AW, Santillán-Galicia M, Tamayo-Mejía F, Bujanos-Muñiz R, García-Ávila CJ, López-Buenfil JÁ (2020) Differential susceptibility of *Bagrada hilaris* (Hemiptera: Pentatomidae) to different species of fungal pathogens. J Econ Entomol 113:50–54

Basnet S, Kamble ST (2018a) RNA interference of the muscle actin gene in bed bugs: exploring injection versus topical application for dsRNA delivery. J Insect Sci 18:5;1–5

Basnet S, Kamble ST (2018b) Knockdown of the chromatin remodeling gene brahma by RNA interference reduces reproductive fitness and lifespan in common bed bug (Hemiptera: Cimicidae). J Med Entomol 55:534–539

Baum JA, Bogaert T, Clinton W, Heck GR, Feldmann P, Ilagan O, Johnson S, Plaetinck G, Munyikwa T, Pleau M, Vaughn T, Roberts J (2007) Control of coleopteran insect pests through RNA interference. Nat Biotechnol 25:1322–1326

Baverstock J, Roy HE, Pell JK (2010) Entomopathogenic fungi and insect behaviour: from unsuspecting hosts to targeted vectors. BioControl 55:89–102

Bloomquist JR (1996) Ion channels as targets for insecticides. Annu Rev Entomol 41:163–190

Britt K, Grant GJ, Wiggins GJ, Stewart SD (2016) Prevalence and localized infection on the entomopathogenic fungus *Beauveria bassiana* on kudzu bug (Hemiptera: Plataspidae) in eastern Tennessee. J Entomol Sci 51:321–324

Cagliari D, Dias NP, Santos EA, Rickes LN, Kremer FS, Farias JR, Lenz G, Galdeano DM, Garcia FRM, Smagghe G, Zotti MJ (2020) First transcriptome of the Neotropical pest *Euschistus heros* (Hemiptera: Pentatomidae) with dissection of its siRNA machinery. Sci Rep 10:4856;1–16

Campos M, Knutson A, Heitholt J, Campos C (2010) Resistance to seed feeding by southern green stink bug, *Nezara viridula* (Linnaeus), in soybean, *Glycine max* (L.) Merrill. Southwest Entomol 35:233–239

Casida JE (2018) Neonicotinoids and other insect nicotinic receptor competitive modulators: progress and prospects. Annu Rev Entomol 63:125–144

Casida JE, Durkin KA (2013) Neuroactive insecticides: targets, selectivity, resistance, and secondary effects. Annu Rev Entomol 58:99–117

Casida JE, Quistad GB (1998) Golden age of insecticide research: past, present or future? Annu Rev Entomol 43:1–16

Castellanos NL, Smagghe G, Sharma R, Oliveira EE, Christiaens O (2019) Liposome encapsulation and EDTA formulation of dsRNA targeting essential genes increase oral RNAi-caused mortality in the Neotropical stink bug *Euschistus heros*. Pest Manag Sci 75:537–548

Cervantes FA, Backus EA, Godfrey L, Akbar W, Clark TL (2016) Characterization of an EPG waveform library for adult *Lygus lineolaris* and *Lygus hesperus* (Hemiptera: Miridae) feeding on cotton squares. Ann Entomol Soc Am 109:684–697

Cervantes FA, Backus EA, Godfrey L, Wallis C, Akbar W, Clark TL, Rojas MG (2017) Correlation of electropenetrography waveforms from *Lygus lineolaris* (Hemiptera: Miridae) feeding on cotton squares with chemical evidence of inducible tannins. J Econ Entomol 110:2068–2075

Cervantes FA, Backus EA, Godfrey L, Rojas MG, Akbar W, Clark TL (2019) Quantitative differences in feeding behavior of *Lygus lineolaris* (Hemiptera: Miridae) on transgenic and nontransgenic cotton. J Econ Entomol 112:1920–1925

Chakrabarty S, Jin M, Wu C, Chakraborty P, Xiao Y (2020) *Bacillus thuringiensis* vegetative insecticidal protein family Vip3A and mode of action against pest Lepidoptera. Pest Manag Sci 76:1612–1617

Chen WB, Lu GQ, Cheng HM, Liu CX, Xiao YT, Xu C, Shen ZC, Wu KM (2017) Transgenic cotton coexpressing Vip3A and Cry1Ac has a broad insecticidal spectrum against lepidopteran pests. J Invertebr Pathol 149:59–65

Chinchilla-Ramírez M, Garzo E, Fereres A, Gavara-Vidal J, ten Broeke CJM, van Loon JJA, Urbaneja A, Pérez-Hedo M (2021) Plant feeding by *Nesidiocoris tenuis*: quantifying its behavioral and mechanical componentes. Biol Control 152:104402;1–10

Christiaens O, Steve Whyard S, Ana M, Vélez AM, Smagghe G (2020) Double-stranded RNA technology to control insect pests: current status and challenges. Frontiers Plant Sci 11:451;1–10

Clark WJ, Harris FA, Maxwell FG, Hartwig EE (1972) Resistance of certain soybean cultivars to bean leaf beetle, striped blister beetle, and bollworm. J Econ Entomol 65:1669–1672

Cline AR, Backus EA (2002) Correlations among AC electronic monitoring waveforms, body postures, and stylet penetration behaviors of *Lygus hesperus* (Hemiptera: Miridae). Environ Entomol 31:538–549

Cooper AM, Silver K, Zhang J, Park Y, Zhu KY (2019) Molecular mechanisms influencing efficiency of RNA interference in insects. Pest Manag Sci 75:18–28

Corrêa-Ferreira BS, Thomazini MJ, Zamataro CE (1991) Effect of *Eutrichopodopsis nitens* Blanchard on adult longevity and reproduction of *Nezara viridula* (L.). Pesq Agropec Bras 26:837–842 (*in Portuguese*)

Corrêa-Ferreira BS, Nunes MC, Uguccioni LD (1998) Occurrence of *Hexacladia smithii* Ashmead parasitoid of *Euschistus heros* (F.) adults in Brazil. An Soc Entomol Brasil 27:495–498 (*in Portuguese*)

Corso IC, Gazzoni DL (1998) Sodium chloride: an insecticide enhancer for controlling pentatomids on soybeans. Pesq Agropec Brasil 33:1563–1571

Davis-Vogel C, Van Allen B, Van Hemert JL, Sethi A, Nelson ME, Sashital DG (2018) Identification and comparison of key RNA interference machinery from western corn rootworm, fall armyworm, and southern green stink bug. PLoS One 13:e0203160;1–26

Dhaliwal GS, Singh R (2004) Host plant resistance to insects: concepts and applications. Panima Publication, New Delhi

Dias NP, Cagliari D, Santos EA, Smagghe G, Jurat-Fuentes JL (2020) Insecticidal gene silencing by RNAi in the Neotropical region. Neotrop Entomol 49:1–11

Diaz-Montano J, Reese JC, Louis J, Campbell LR, Schapaugh WT (2007) Feeding behavior by the soybean aphid (Hemiptera: Aphididae) on resistant and susceptible soybean genotypes. J Econ Entomol 100:984–989

Douglas AE (1998) Nutritional interactions in insect-microbial symbioses: aphids and their symbiotic bacteria *Buchnera*. Annu Rev Entomol 43:17–37

Duyn JWV, Turnipseed SG, Maxwell JD (1971) Resistance in soybean to the Mexican bean beetle. I. Source of resistance. Crop Sci 11:572–573

Eger JE (1981) Oviposition site preference and related aspects of the parasitism of *Brochymena* spp. by *Trichopoda plumipes* in College Station, Texas. Environ Entomol 10:468–471

Eghrari K, Brito AH, Baldassi A, Balbuena TS, Fernandes OA, Môro GV (2019) Homozygosis of Bt locus increases Bt protein expression and the control of *Spodoptera frugiperda* (Lepidoptera: Noctuidae) in maize hybrids. Crop Prot 124:104871

Gerolt P (1969) Mode of entry of contact insecticides. J Insect Physiol 15:563–580
Ghaffar MBBA, Pritchard J, Ford-Lloyd B (2011) Brown planthopper (*N. lugens* Stal) feeding behaviour on rice germplasm as an indicator of resistance. PLoS One 6:e22137;1–113
Ghosh SKB, Hunter WB, Park AL, Gundersen-Rindal DE (2017) Double strand RNA delivery system for plant-sap-feeding insects. PLoS One 12:e0171861;1–19
Gouli V, Gouli S, Skinner M, Hamilton G, Su Kim J, Parker BL (2012) Virulence of select entomopathogenic fungi to the brown marmorated stink bug, *Halyomorpha halys* (Stål) (Heteroptera: Pentatomidae). Pest Manag Sci 68:155–157
Gurusamy D, Howell JL, Chereddy SCRR, Koo J, Palli SR (2020) Transport of orally delivered dsRNA in southern green stink bug, *Nezara viridula*. Arch Insect Biochem Physiol 104;e21692;1–18
Hirose E, Panizzi AR, Cattelan AJ (2006) Potential use of antibiotic to improve performance of laboratory-reared *Nezara viridula* (L.) (Heteroptera: Pentatomidae). Neotrop Entomol 35:279–281
Hori K (2000) Possible causes of disease symptons resulting from the feeding of phytophagous Heteroptera. In: Schaefer CW, Panizzi AR (eds) Heteroptera of economic importance. CRC Press, Boca Raton, pp 11–35
Howell JL, Mogilicherla K, Gurusamy D, Palli SR (2020) Development of RNAi methods to control the harlequin bug, *Murgantia histrionica*. Arch Insect Biochem Physiol 104:e21690;1–14
Huvenne H, Smagghe G (2010) Mechanisms of dsRNA uptake in insects and potential of RNAi for pest control: a review. J Insect Physiol 56:227–235
Jacobson AL, Kennedy GG (2014) Electrical penetration graph studies to investigate the effects of cyantraniliprole on feeding behavior of *Myzus persicae* (Hemiptera: Aphididae) on *Capsicum annuum*. Pest Manag Sci 70:836–840
Joga MR, Zotti MJ, Smagghe G, Christiaens O (2016) RNAi efficiency, systemic properties, and novel delivery methods for pest insect control: what we know so far. Frontiers Physiol 7:553;1–14
Jones WA, Sullivan JM (1979) Soybean resistance to the southern green stink bug, *Nezara viridula*. J Econ Entomol 72:628–632
Kamminga KL, Davis JA, Stock SP, Ritcher AR (2012) First report of a mermithid nematode infecting *Piezodorus guildinii* and *Acrosternum hilare* (Hemiptera: Pentatomidae) in the United States. Fla Entomol 95:214–217
Karabörklü S, Azizoglu U, Azizoglu ZB (2018) Recombinant entomopathogenic agents: a review of biotechnological approaches to pest insect control. World J Microb Biotechnol 34:14;1–12
Kariñho-Betancourt E, Núñez-Farfán J (2015) Evolution of resistance and tolerance to herbivores: testing the trade-off hypothesis. PeerJ 3:e789;1–15
Koch KG, Donze-Reiner T, Baird LM, Louis J, Amundsen K, Sarath G, Bradshaw JD, Heng-Moss T (2018) Evaluation of greenbug and yellow sugarcane aphid feeding behavior on resistant and susceptible switchgrass cultivars. Bioenergy Res 11:480–490
Latham JR, Love M, Hilbeck A (2017) The distinct properties of natural and GM cry insecticidal proteins. Biotechnol Genet Eng 33:62–96
Li W, Wyckhuys KAG, Wu K (2016) Does feeding behavior of a zoophytophagous mirid differ between host plant and insect prey items? Arthropod Plant Interact 10:79–86
Liu J, Poinar GO Jr, Berry RE (2000) Control of insect pests with entomopathogenic nematodes: the impact of molecular biology and phylogenetic reconstruction. Annu Rev Entomol 45:287–306
Liu H, Skinner M, Parker BL, Brownbridge M (2002) Pathogenicity of *Beauveria bassiana, Metarhizium anisopliae* (Deuteromycotina: Hyphomycetes), and other entomopathogenic fungi against *Lygus lineolaris* (Hemiptera: Miridae). J Econ Entomol 95:675–681
Liu XD, Zhai BP, Zhang XX, Zong MJ (2005) Impact of transgenic cotton plants on a non-target pest, *Aphis gossypii* Glover. Ecol Entomol 30:307–315
Lourenção AL, Miranda MAC, Nagai V (1987) Soybean resistance to insects: VII. Evaluation of stink bug damage to cultivars and lines. Bragantia 46:45–57 (*in Portuguese*)

Lu Z, Dong S, Li C, Li L, Yu Y, Men X, Yin S (2020) Sublethal and transgenerational effects of dinotefuran on biological parameters and behavioural traits of the mirid bug *Apolygus lucorum*. Sci Rep 10:226;1–8

Lucini T, Panizzi AR (2016) Waveform characterization of the soybean stem feeder *Edessa meditabunda* (F.) (Hemiptera: Heteroptera: Pentatomidae): overcoming the challenge of wiring pentatomids for EPG. Entomol Exp Appl 158:118–132

Lucini T, Panizzi AR (2017a) Feeding behavior of the stink bug *Dichelops melacanthus* Dallas on corn seedlings: an EPG analysis at multiple input impedances and histology correlation. Ann Entomol Soc Am 110:160–171

Lucini T, Panizzi AR (2017b) Probing behavior of *Dichelops furcatus* (F.) (Heteroptera: Pentatomidae) on wheat plants characterized by electropenetrography (EPG) and histological studies. J Insect Sci 17:65;1–15

Lucini T, Panizzi AR (2018) Electropenetrography monitoring of the neotropical brown-stink bug (Hemiptera: Pentatomidae) on soybean pods: an electrical penetration graph-histology analysis. J Insect Sci 18:5;1–14

Lucini T, Panizzi AR, Backus EA (2016) Characterization of an EPG waveform library for red-banded stink bug, *Piezodorus guildinii* (Hemiptera: Pentatomidae), on soybean plants. Ann Entomol Soc Am 109:198–210

Lucini T, Panizzi AR, Dios RVP (2020) Tachinid fly parasitism and phenology of the Neotropical red-shouldered stink bug, *Thyanta perditor* (F.) (Heteroptera: Pentatomidae), on the wild host plant, *Bidens pilosa* L. (Asteraceae). Neotrop Entomol 49:98–107

Luedders VD, Dickerson WA (1977) Resistance of selected soybean genotypes and segregating populations to cabbage looper feeding. Crop Sci 17:395–397

Marques LH, Santos AC, Castro BA, Moscardini VF, Rossetto J, Silva OAN, Zobiole LHS, Valverde-Garcia P, Babcock JM, Storer NP, Rule DM, Fernandes AO (2017) Field evaluation of soybean transgenic event DAS-81419-2 expressing Cry1F and Cry1Ac proteins for the control of secondary lepidopteran pests in Brazil. Crop Prot 96:109–115

Martin PA, Hirose E, Aldrich JR (2007) Toxicity of *Chromobacterium subtsugae* to southern green stink bug (Heteroptera: Pentatomidae) and corn rootworm (Coleoptera: Chrysomelidae). J Econ Entomol 100:680–684

Maxwell FG, Jennings PR (1980) Breeding plants resistant to insects. Wiley, New York

McPherson JE, McPherson RM (2000) Stink bugs of economic importance in America north of Mexico. CRC Press, Boca Raton

McPherson RM, Buss GR, Roberts PM (2007) Assessing stink bug resistance in soybean breeding lines containing genes from germplasm IAC-100. J Econ Entomol 100:1456–1463

Miao J, Han BY, Zhang QH (2014) Probing behavior of *Empoasca vitis* (Homoptera: Cicadellidae) on resistant and susceptible cultivars of tea plants. J Insect Sci 14:223;1–6

Mitchell PL, Cooke SB, Smaniotto LF (2018) Probing behavior of *Nezara viridula* on soybean: characterization and comparison of electrical penetration graph (EPG) waveforms on vegetative and reproductive plant structures. J Agric Urban Entomol 34:19–43

Mogilicherla K, Howell JL, Palli SR (2018) Improving RNAi in the brown marmorated stink bug: identification of target genes and reference genes for RT-qPCR. Sci Rep 8:3720;1–9

Moscardi F (1999) Assessment of the application of baculoviruses for control of Lepidoptera. Annu Rev Entomol 44:257–289

Mustafa T, Alvarez JM, Munyaneza JE (2015) Effect of cyantraniliprole on probing behavior of the potato psyllid (Hemiptera: Triozidae) as measured by the electrical penetration graph technique. J Econ Entomol 108:2529–2535

Niva CC, Panizzi AR (1996) Effects of sodium chloride on the behavior of *Nezara viridula* L. (Heteroptera: Pentatomidae) on soybean pod. An Soc Entomol Brasil 25:252–257 (*in Portuguese*)

Núñez-Farfan J, Fornoni J, Valverde PL (2007) The evolution of resistance and tolerance to herbivores. Annu Rev Ecol Evol Syst 38:541–566

O'Brien RD (1966) Mode of action of insecticides. Annu Rev Entomol 11:369–402

Orians CM, Ward D (2010) Evolution of plant defenses in nonindigenous environments. Annu Rev Entomol 55:439–459

Painter RH (1951) Insect resistance in crop plants. McMillan, New York

Panda N (1979) Principles of host-plant resistance to insect pests. Hindustan Publication, New Delhi

Panizzi AR, da Silva JJ (2010) New records of pentatomids as hosts of *Hexacladia smithii* Ashmead (Hymenoptera: Encyrtidae) in Southern Brazil. Neotrop Entomol 39:678–679

Panizzi AR, Slansky F Jr (1985) Review of phytophagous pentatomids associated with soybean in the Americas. Fla Entomol 68:184–214

Panizzi AR, McPherson JE, James DG, Javahery M, McPherson RM (2000) Economic importance of stink bugs (Pentatomidae). In: Schaefer CW, Panizzi AR (eds) Heteroptera of economic importance. CRC Press, Boca Raton, pp 421–474

Parys KA, Portilla M (2020) Effectiveness of *Beauveria bassiana* against *Piezodorus guildinii* (Hemiptera: Pentatomidae), a key pest of soybeans in the neotropics. Biocontrol Sci Tech 30:451–461

Pinheiro DH, Moreira RO, Leite NA, Redoan AN, Xavier AS, Barros BA, Carneiro NP (2020) Suitable reference genes for RT-qPCR analysis in *Dichelops melacanthus* (Hemiptera: Pentatomidae). Mol Biol Rep 47:4989–5000

Prado SS, Zucchi TD (2012) Host-symbiont interactions for potentially managing heteropteran pests. Psyche 2012:269473;1–9

Price DR, Gatehouse JA (2008) RNAi-mediated crop protection against insects. Trends Biotechnol 26:393–400

Qaim M, Zilberman D (2003) Yield effects of genetically modified crops in developing countries. Science 299:900–902

Rangasamy M, McAuslane HJ, Backus EA, Cherry AH (2015) Differential probing behavior of *Blissus insularis* (Hemiptera: Blissidae) on resistant and susceptible St. Augustinegrasses. J Econ Entomol 108:780–788

Resquín-Romero G, Cabral-Antúnez C, Sarubbi-Orue H, Garrido-Jurado I, Valverde-García P, Schade M, Butt TM (2020) Virulence of *Metarhizium brunneum* (Ascomycota: Hypocreales) strains against stinkbugs *Euschistus heros* and *Dichelops furcatus* (Hemiptera: Pentatomidae). J Econ Entomol 113:2540–2545

Ribeiro A, Castiglioni E (2009) Fluctuaciones de poblacionales de *Piezodorus guildinii* (Westwood) (Hemiptera: Pentatomidae) en soja y alfafa em Paysandú, Uruguay. Agrociencia Uruguaya 23:32–36

Riga M, Denecke S, Livadaras I, Geibel S, Nauen R, Vontas J (2020) Development of efficient RNAi in *Nezara viridula* for use in insecticide target discovery. Arch Insect Biochem Physiol 103:e21650;1–9

Rivera KL, Mitchell PL (2020) Feeding behavior of *Nezara viridula* first instars: EPG analysis and effect of food availability on subsequent development. J Agric Urban Entomol 36:47–63

Roan CC, Hopkins TL (1961) Mode of action of insecticides. Annu Rev Entomol 6:333–346

Rossetto CJ, Gallo PB, Razera LF, Bortoletto N, Igue T, Medina PF, Tisseli Filho O, Aquilera V, Veiga RFA, Pinheiro JB (1995) Mechanisms of resistance to stink bug complex in the soybean cultivar IAC-100. An Soc Entomol Brasil 24:517–522

Sadasivam S, Thayumanavan B (2003) Molecular host plant resistance to pests. Marcel Dekker, New York

Sánchez JA, Lacasa A (2008) Impact of the zoophytophagous plant bug *Nesidiocoris tenuis* (Heteroptera: Miridae) on tomato yield. J Econ Entomol 101:1864–1870

Schaefer CW, Panizzi AR (2000) Heteroptera of economic importance. CRC Press, Boca Raton

Seiter NJ, Grabke A, Greene JK, Kerrigan JL, Reay-Jones FPF (2014) *Beauveria bassiana* is a pathogen of *Megacopta cribraria* (Hemiptera: Plataspidae) in South Carolina. J Entomol Sci 49:326–330

Serikawa RH, Backus EA, Rogers ME (2012) Effects of soil-applied imidacloprid on Asian citrus psyllid (Hemiptera: Psyllidae) feeding behavior. J Econ Entomol 105:1492–1502

Serteyn L, Ponnet L, Backus EA, Francis F (2020a) Characterization of electropenetrography waveforms for the invasive heteropteran pest, *Halyomorpha halys*, on *Vicia faba* leaves. Arthropod Plant Interact 14:113–126

Serteyn L, Ponnet L, Saive M, Fauconnier M-L, Francis F (2020b) Changes of feeding behavior and salivary proteome of brown marmorated stink bug when exposed to insect-induced plant defenses. Arthropod Plant Interact 14:101–112

Sharma R, Christiaens O, Taning CNT, Smagghe G (2021) RNAi-mediated mortality in Southern green stinkbug *Nezara viridula* by oral delivery of dsRNA. Pest Manage Sci 77:77–84

Silva JPGF, Baldin ELL, Souza ES, Canassa VF, Lourenção AL (2013) Characterization of antibiosis to the redbanded stink bug *Piezodorus guildinii* (Hemiptera: Pentatomidae) in soybean entries. J Pest Sci 86:649–657

Silva JPGF, Baldin ELL, Canassa VF, Souza ES, Lourenção AL (2014) Assessing antixenosis of soybean entries against *Piezodorus guildinii* (Hemiptera: Pentatomidae). Arthrop Plant Interact 8:349–359

Smaniotto LF, Panizzi AR (2015) Interactions of selected species of stink bugs (Hemiptera: Heteroptera: Pentatomidae) from leguminous crops with plants in the neotropics. Fla Entomol 98:7–17

Smith CM (1989) Plant resistance to insects. A fundamental approach. Wiley, New York

Smith CM, Clement SL (2012) Molecular bases of plant resistance to arthropods. Annu Rev Entomol 57:309–328

Smith CM, Khan ZR, Pathak MD (1994) Techniques for evaluating insect resistance in crop plants. CRC Press, Boca Raton

Snelling RO (1941) Resistance of plants to insect attack. Bot Rev 7:543–586

Spiller NJ, Koenders L, Tjallingii WF (1990) Xylem ingestion by aphids – a strategy for maintaining water balance. Entomol Exp Appl 55:101–104

Stout MJ (2013) Reevaluating the conceptual framework for applied research on host-plant resistance. Insect Sci 20:263–272

Stubbins FL, Agudelo P, Reay-Jones FPF, Greene JK (2015) First report of a mermithid nematode infecting the invasive *Megacopta cribraria* (Hemiptera: Plataspidae) in the United States. J Invertebr Pathol 127:35–37

Stubbins FL, Agudelo P, Reay-Jones FPF, Greene JK (2016) *Agamermis* (Nematoda: Mermithidae) infection in South Carolina agricultural pests. J Nematol 48:290–296

Stubbins FL, Mitchell PL, Turnbull MW, Reay-Jones FPF, Greene JK (2017) Mouthpart morphology and feeding behavior of the invasive kudzu bug, *Megacopta cribraria* (Hemiptera: Plataspidae). Invertebr Biol 136:309–320

Sun YP (1968) Dynamics of insect toxicology-a mathematical and graphical evaluation of the relationship between insect toxicity and rates of penetration and detoxication of insecticides. J Econ Entomol 61:949–955

Sun M, Voorrips RE, Steenhuis-Broers G, van't Westende W, Vosman B (2018a) Reduced phloem uptake of *Myzus persicae* on an aphid resistant pepper accession. BMC Plant Biol 18:138;1–14

Sun Z, Yan F, Wang MQ (2018b) Transgenic expression of Bt in rice does not affect feeding behavior and population density of the brown planthopper, *Nilaparvata lugens* Stål (Hemiptera: Delphacidae). Entomol Gen 37:35–45

Taiz L, Zeiger E (2004) Plant physiology, 3rd edn. Sinauer Associates Inc., Sunderland

Tanada Y (1959) Microbiol control of insects. Annu Rev Entomol 4:277–302

Tariq K, Noor M, Backus EA, Hussain A, Ali A, Peng W, Zhang H (2017) The toxicity of flonicamid to cotton leafhopper, *Amrasca biguttula* (Ishida), is by disruption of ingestion: an electropenetrography study. Pest Manag Sci 73:1661–1669

Todd JC, Rouf Mian MA, Backus EA, Finer JJ, Redinbaugh MG (2016) Feeding behavior of soybean aphid (Hemiptera: Aphididae) biotype 2 on resistant and susceptible soybean. J Econ Entomol 109:426–433

Tozlu E, Saruhan I, Tozlu G, Kotan R, Dadaşoğlu F, Tekiner N (2019) Potentials of some entomopathogens against the brown marmorated stink bug, *Halyomorpha halys* (Stål, 1855) (Hemiptera: Pentatomidae). Egypt J Biol Pest Control 29:76;1–8

Tuelher ES, Backus EA, Cervantes F, Oliveira EE (2020) Quantifying *Lygus lineolaris* stylet probing behavior and associated damage to cotton leaf terminals. J Pest Sci 93:663–677
Walker WB, Allen ML (2010) Expression and RNA interference of salivary polygalacturonase genes in the tarnished plant bug, *Lygus lineolaris*. J Insect Sci 10:173;1–13
Walker WB, Allen ML (2011) RNA interference-mediated knockdown of IAP in *Lygus lineolaris* induces mortality in adult and pre-adult life stages. Entomol Exp Appl 138:83–92
Wang YN, Kel KO, Li YH, Han LZ, Liu YM, Hua HX, Peng YF (2016) Comparison of three transgenic Bt rice lines for insecticidal protein expression and resistance against a target pest, *Chilo suppressalis* (Lepidoptera: Crambidae). Insect Sci 23:78–87
Waqas MS, Qian L, Shoaib AAZ, Cheng X, Zhang Q, Elabasy ASS, Shi Z (2019) Lethal and sublethal effects of neonicotinoid insecticides on the adults of *Phenacoccus solenopsis* (Hemiptera: Pseudococcidae) on tomato plants. J Econ Entomol 112:1314–1321
Williamson C, von Wechmar MB (1995) The effect of two viruses on the metamorphosis, fecundity, and longevity of the green stinkbug, *Nezara viridula*. J Invert Pathol 65:174–178
Wilson RC, Doudna JA (2013) Molecular mechanisms of RNA interference. Annu Rev Biophys 42:217–239
Xue K, Wang XY, Huang CH, Wang RJ, Liu B, Yan FM, Xu CR (2009) Stylet penetration behaviors of the cotton aphid *Aphis gossypii* on transgenic Bt cotton. Insect Sci 16:137–146
Yencho GC, Cohen MB, Byrne PF (2000) Applications of tagging and mapping insect resistance loci in plants. Annu Rev Entomol 45:393–422
Yin HD, Wang XY, Xue K, Huang CH, Wang RJ, Yan FM, Xu CR (2010) Impacts of transgenic Bt cotton on the stylet penetration behaviors of *Bemisia tabaci* biotype B: evidence from laboratory experiments. Insect Sci 17:344–352
Zavala JA, Mazza CA, Dillon FM, Chludil HD, Ballaré CL (2015) Soybean resistance to stink bugs (*Nezara viridula* and *Piezodorus guildinii*) increases with exposure to solar UV-B radiation and correlates with isoflavonoid content in pods under field conditions. Plant Cell Environ 38:920–928
Zhu KY, Palli SR (2020) Mechanisms, applications, and challenges of insect RNA interference. Annu Rev Entomol 65:293–311
Zlotkin E (1999) The insect voltage-gated sodium channel as target of insecticides. Annu Rev Entomol 44:429–455

Chapter 8
Perspectives on the Use of EPG in Electronic Monitoring of Phytophagous True Bugs

Contents

Abstract In this concluding chapter, we discuss the value of EPG in electronic monitoring of all activities of phytophagous true bugs. Although the majority of studies have concentrated on feeding behavior, EPG is a tool with the potential to reveal details of any activity that generates an electrical connection between the bug and the plant, which expands its application. We also explore how EPG may help to interpret and predict plant damage based on the waveforms produced, focusing on host plants, feeding sites, and duration of feeding. Finally, we consider how information generated using the EPG can enhance strategies/tactics of integrated pest management (IPM) programs.

Keywords Future of EPG · Feeding behavior · Non-feeding behaviors · Plant damage · IPM

8.1 Introduction

Since the beginning of research work using electronic devices (we will focus on the electrical penetration graph or electropenetrography – EPG) to monitor the feeding activity of insects in the 1960s (see Sect. 1.3 in Chap. 1), a long and winding way has been traveled.

A. R. Panizzi et al., *Electronic Monitoring of Feeding Behavior of Phytophagous True Bugs (Heteroptera)*, Entomology in Focus 6,
https://doi.org/10.1007/978-3-030-64674-5_8

New and more sophisticated apparatus were developed over time (for details, see Sect. 1.3.3 in Chap. 1). This new equipment allowed the generation of more accurate electronic waveforms, not only for the many activities related to feeding but for other activities as well, as we will discuss ahead. For example, egg laying by the pentatomid *Diceraeus (Dichelops) melacanthus* (Dallas) on maize stems generated a unique waveform, each peak representing the deposition of a single egg (Lucini and Panizzi 2017). This waveform allows one to count the total number of eggs deposited in a single egg mass, without the need to examine the egg mass per se. Another example is the generation of waveforms that present a chaotic pattern of multiple peaks in sequence that indicate movement of the bug on the plant surface. By examining that waveform, one can estimate the "wandering time" of the bug. A flat line indicates the bug standing still on the plant surface and so on. These various behaviors, and the waveforms generated, will be discussed and illustrated in greater detail ahead.

For now, we simply want to call attention to the potential of EPG as an invaluable tool to reveal the main activities of true bugs. By monitoring the plant/bug interface, EPG generates electrical signals which are all recorded and converted to waveforms for later examination and interpretation. This, of course, is a "do and learn experience," and it takes time to fully understand the process. Up to this point, however, we can be sure that EPG certainly is, as stated by Lucini and Panizzi (2018), a breakthrough tool revealing true bug feeding on plants.

8.2 Importance of EPG Electronic Monitoring to Reveal Details of Feeding Behavior

Feeding behavior of hemipteroids is perhaps one of the most complex and sophisticated activities performed by insects. In the case of true bugs, feeding occurs inside the food source; thus, it cannot be observed in detail visually or using video recording devices, as one can do, for example, with chewing insects.

For true bugs, some aspects of the feeding behavior on plants have been observed and described over the years using other methods. In general, the information published covers stylet penetration, feeding sites, external deposits of gelling saliva (called flanges), stylet sheath formation internally, and feeding frequency and duration (e.g., Miles 1972; Bowling 1980; Simmons and Yeargan 1988; Panizzi 1995; Panizzi et al. 1995; Hori 2000; Depieri and Panizzi 2011). These parameters were mostly recorded through visual observation and helped to characterize these traits. However, despite these valuable observations, they have some limitations. Visual observations are, in general, carried out in a narrow timeframe, which is perfectly understandable. Because of that, unexpected or uncommon (yet sometimes important) behaviors may be missed. Moreover, they are dependent on the skill of the observer, and this varies from one person to another and is, therefore, susceptible to subjectivity.

The use of tools such as EPG to monitor the feeding behavior of true bugs enables us to expand our understanding of the process and add what we can call a "degree of certainty." As activities are recorded over a long time span (e.g., 15 h), by analyzing the waveforms generated, one can calculate precisely the duration of the different behaviors and construct a "time budget" allocated to each activity. For example, let us consider all activities (not only feeding) of the stink bug *Diceraeus (Dichelops) furcatus* (F.) on several food sources that were recorded using EPG (Fig. 8.1).

Looking at Fig. 8.1, the first thing to note is the great amount of time that the bugs spent resting or walking on the plant surface (Np and Z combined waveforms). These waveforms account for roughly 60–80% of the total recording time. This suggests that most of the time, bugs do not feed, and this is the opposite of what is generally thought. For instance, because of the great damage inflicted to seeds or fruits, growers believe that bugs feed continuously. Feeding behaviors (i.e., those performed with stylets inserted in the plant tissue) account for the rest of the activi-

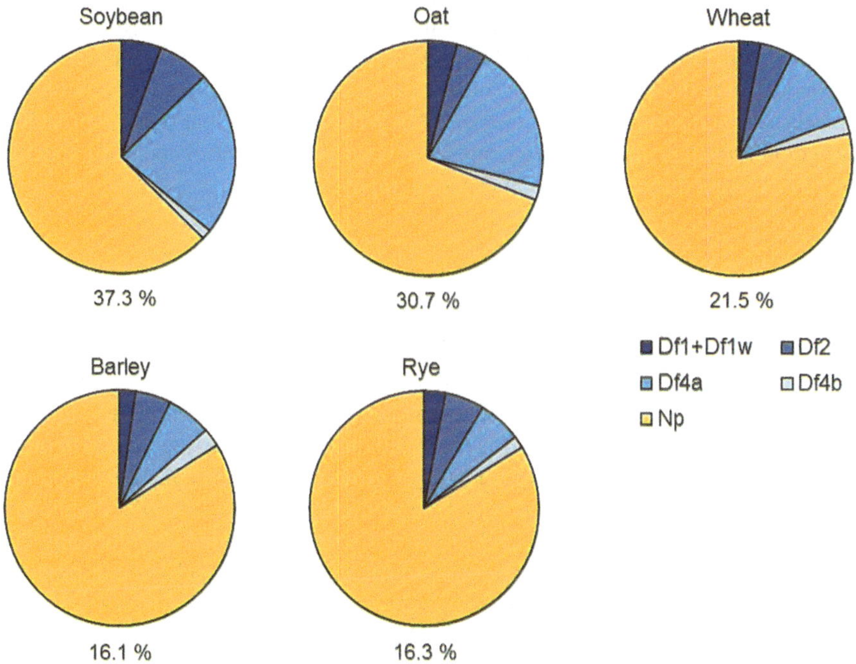

Fig. 8.1 Total duration of each waveform (TWD; pie slices) recorded during EPG studies of the stink bug *Diceraeus (Dichelops) furcatus* on immature soybean pods and immature seed heads of oat, barley, wheat, and rye. The value beneath each pie chart represents the total time spent with stylets inserted, expressed as a percentage of total activities. Df1 = combined pathway waveforms (Df1a, Df1b) = stylet penetration into the plant tissue; Df1w = stylet withdrawal from tissue; Df2 = xylem sap ingestion; Df4a = cell laceration, enzymatic maceration of seed endosperm; Df4b = ingestion of macerated seed endosperm; Np = insects resting/walking on plant surface; i.e., all waveforms not involving stylet insertion in plant tissue (Np, Z). (Source: adapted from Lucini and Panizzi 2020)

ties, which vary according to the different food sources but never account for more than 40% of the recording time.

For example, bugs spent a larger percentage of time on combined feeding activities (sum of the durations of waveforms Df1, Df1w, Df2, Df4a, and Df4b – the meaning of each waveform is explained in the legend of Fig. 8.1) on soybean and oat than on barley and rye plants, while the duration on wheat was intermediate (Fig. 8.1). Detailed feeding activities include, for instance, cell laceration and enzymatic maceration of seed endosperm (waveform Df4a) which constitutes the second longest time-consuming activity. The time budget may be constructed in this manner, until 100% of the recording time is complete.

In addition, waveforms correlated with histological studies reveal the different feeding sites in greater detail (see Chap. 6 for more details). Therefore, the coupling of electronic equipment with histology adds a tremendous amount of specific information on the feeding activities. Comparing this with the information acquired solely by visual observation, one can see the great advantage of using electronic equipment to register activities performed by the bugs. However, we do not want readers to think that we are excluding visual observation or video recording devices as a way to get information on the bugs' behavior. In fact, insect visual observation in the lab or in the field is, among other things, a rewarding, pleasant activity that reveals so much about them (Bernays 2020). What is clear is that these approaches are complementary; many waveforms have been visually correlated with specific behaviors using direct observation or videography.

8.3 Importance of EPG Electronic Monitoring to Reveal Details of Other Behaviors

Although electronic monitoring devices are widely used to determine the feeding behavior of insects, some waveforms produced during recording are related to other activities. Often, when talking to students or growers, questions like these pop up: How long do the bugs feed? Do they feed at night? Do they "sleep"? People are curious to know these answers, and most of the time we ended up "guessing" in response. We believe that electronic monitoring devices such as EPG can produce more precise answers to these types of questions.

As the bugs are wired and connected to the EPG monitor and the recording starts, what we see in the monitor is depicted in Fig. 8.2. A continuous line with very small undulations which represents the bug standing still (upper box) is observed. This waveform presents low amplitude with slight variation in voltage and is the baseline of the recording. As the bug starts moving on the plant surface, a waveform with numerous variable spikes/peaks is produced, which is easily identified. This waveform has a chaotic pattern; i.e., peaks and valleys occur interspersed without uniformity of amplitude or frequency (lower box). These signals are caused by electrical contacts of tarsal claws while scratching the plant surface.

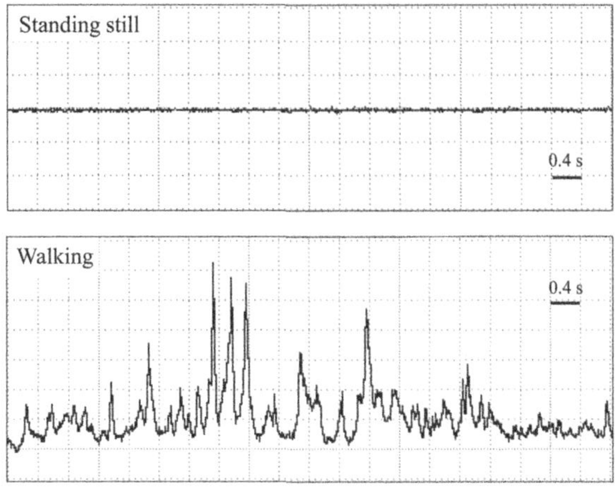

Fig. 8.2 Waveforms recorded using EPG from the stink bug *Euschistus heros* on soybean pods (50 mV AC applied voltage, 10^8 Ohms input impedance) describing two different activities not related to feeding (standing still and walking)

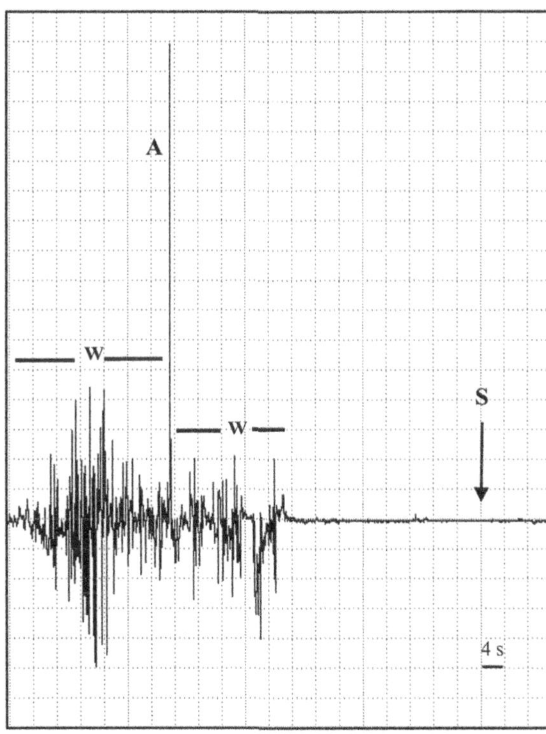

Fig. 8.3 Detail of the non-feeding waveforms (A, S and W) (50 mV AC applied voltage, 10^9 Ohms input impedance) recorded from *Lygus* spp. A = insect touches the cotton surface with antennal tips (antennation behavior); W = insect walking on cotton square; S = insect standing still on cotton square. (Source: Cervantes et al. 2016)

Another non-feeding activity is antennation, which has been recorded using EPG only for *Lygus* species feeding on pinhead cotton squares. This brief behavior is detected only in recordings made at high input impedance levels (over 10^8 Ohms). It was visually correlated with the bug performing single or multiple taps of the antennal tips on the cotton square surface. A single antennal touch is represented by a single, positive, high amplitude spike (Fig. 8.3) (Cervantes et al. 2016).

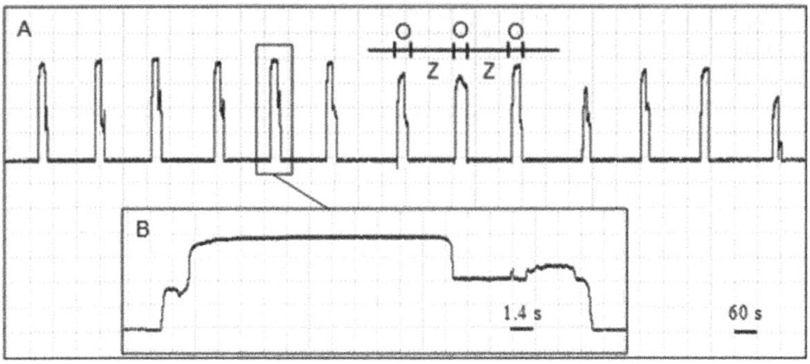

Fig. 8.4 Detail of the waveform O recorded from *Diceraeus (Dichelops) melacanthus* on stem of maize seedlings (50 mV AC applied voltage, 10^6 Ohms input impedance). Multiple O waveforms, representing egg deposition, interspersed with Z waves (resting) (**a**). Detail of a single waveform O (**b**). (Source: Lucini and Panizzi 2017)

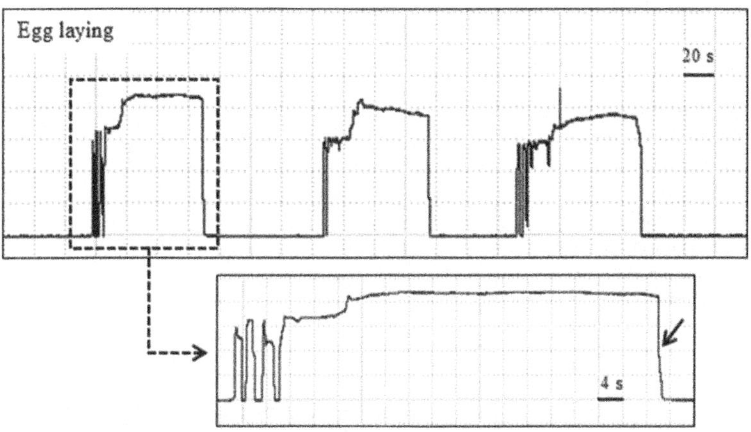

Fig. 8.5 Egg laying waveform recorded using EPG from the stink bug *Euschistus heros* on soybean pods (50 mV AC applied voltage, 10^7 Ohms input impedance) showing three egg laying waveforms (upper box) and a single, detailed egg laying waveform (bottom box)

When female insects are used in EPG studies with true bugs, egg laying may occur during the recordings. In our studies, females laid eggs on two occasions. In Fig. 8.4, we show details of a waveform of *D. melacanthus* ovipositing on a maize plant stem, each one displaying as a long and flat plateau waveform corresponding to deposition of one egg (waveform O) interspersed with Z waves (bug at rest).

In Fig. 8.5, we show a waveform of the stink bug *Euschistus heros* (F.) laying eggs on soybean pods. As in *D. furcatus*, each long flat peak represents one egg deposited, followed by a valley separating it from the following egg (upper box). As egg laying stops, the waveform turns into a more or less flat line (resting waveform) with a sharp drop in voltage as the female stops touching the plant surface with its genitalia (lower box; indicated by the arrow).

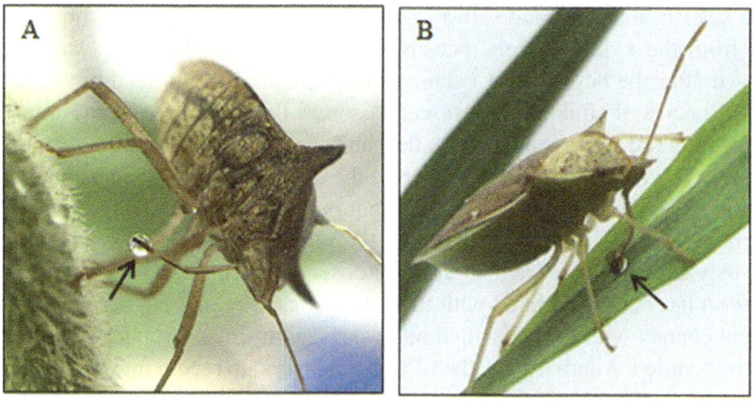

Fig. 8.6 Adult of the pentatomids *Euschistus heros* (**a**) and *Diceraeus* (*Dichelops*) *furcatus* (**b**) carrying a droplet (indicated by the arrows) of food/saliva regurgitate on soybean pod and wheat leaf, respectively, after a feeding session, probably on xylem vessels. (Photos **a** by T Lucini and **b** by AR Panizzi)

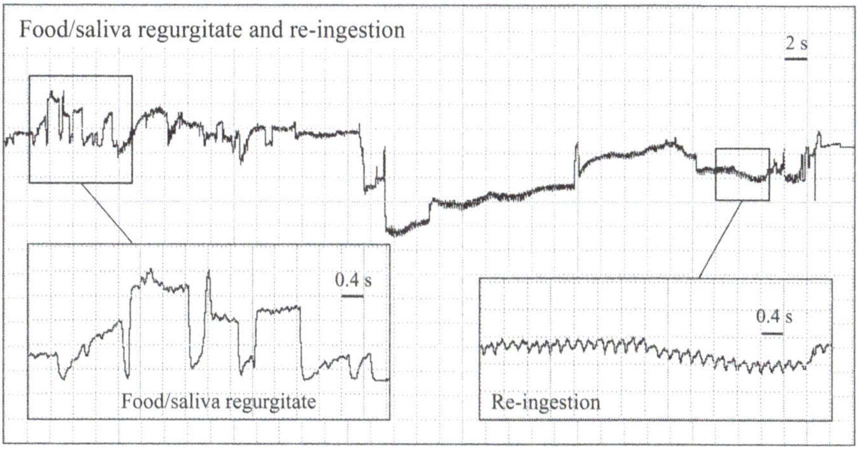

Fig. 8.7 Overview of waveforms recorded using EPG from the stink bug *Euschistus heros* on soybean pods (50 mV AC applied voltage, 10^7 Ohms input impedance) describing a less conspicuous activity related to feeding (food/saliva regurgitate and re-ingestion). See text for explanation

One other behavior, less conspicuous, that was observed during our recording of the stink bug *E. heros* is food/saliva regurgitation and probable later re-ingestion. Small droplets of food/saliva are produced at the tip of the labium (Fig. 8.6a) and subsequently deposited on the plant surface. For each droplet deposited, a peak with a somewhat large plateau appears in the waveform followed by a valley (Fig. 8.7 – detail in the bottom left box). Bugs drag the droplets for short distances and may re-ingest the food/saliva, producing a waveform with a continuous undulating line (Fig. 8.7 – detail in the bottom right box).

We suspect that the food/saliva regurgitate is a result of excess liquid (water) taken from the xylem vessels, because for *E. heros* this behavior was primarily observed after the bugs fed on xylem. This suspicion is reinforced by the fact that another species of stink bug, *D. furcatus*, shows the same behavior (Panizzi et al. 2015) (Fig. 8.6b) after feeding on leaflets and stems of wheat seedlings, most likely on xylem. The reason why re-ingestion of the droplets occurs remains unclear.

Other behaviors, for example, cleaning mouthparts using the tarsi, and the rapid movements of tarsi rubbed one against the other, which the bugs commonly do when they stop moving, are also visually observed during EPG recordings, but no waveform has been correlated with those activities. The moving body parts are not in direct contact with the electrified plant surface; this may explain why no voltage spike is recorded. Alternatively, the EPG settings used to record may not be properly set to capture and register these behaviors. In any case, other devices and/or visual observation must be used. Video recordings have been widely used to register these and other behaviors, not only of insects but of animals in general (Wratten 1994).

8.4 Plant Damage and EPG

The damage inflicted to plants by true bugs was illustrated in Chap. 1 (Sect. 1.6) and covered in detail Chap. 3 (Sect. 3.3), including the relationship of damage to feeding sites and feeding strategies. Therefore, here we will only briefly highlight how EPG may help to interpret the damage resulting from feeding activity based on the waveforms produced and their duration.

Let us analyze, for example, the feeding of true bugs on two main feeding sites: the vascular tissues (xylem/phloem), and the seed endosperm, relating the duration of feeding sessions to the resulting plant damage caused by different species of stink bugs. When feeding on xylem/phloem tissues, bugs usually remain in feeding sessions for a relatively long time. For example, for the stem feeder specialist *Edessa meditabunda* (F.), feeding events last 1 h on xylem and for about 1.5 h in the phloem (Table 8.1). These events cause the usual damage that we see on host plants, such as dark spots on stems of soybean (Panizzi and Machado-Neto 1992), and wilting and death of growing tips, usually on leafy vegetables such as potato, chicory, and lettuce (Rizzo 1971; Krinski and Pelissari 2012; Krinski 2013).

When considering species of bugs that prefer seeds/fruits such as *Nezara viridula* (L.) and *Piezodorus guildinii* (Westwood), the durations of the feeding sessions on the different sites change. They feed longer (over 80 min; this time includes preparation of the tissue plus ingestion of the cell contents) on the soybean seed endosperm and for less time (ca. 40 min) on the xylem from different plant structures (Table 8.1). The resulting damage appears mostly on the seeds that develop abnormally and sometimes are completely damaged (e.g., Todd and Turnipseed 1974).

The pentatomids *D. furcatus* and *D. melacanthus* may feed on wheat and maize seedlings, respectively. Because they have preference for reproductive structures of their hosts, on seedlings they will ingest from xylem using the salivary sheath strat-

Table 8.1 Mean (± SE) duration (min) of feeding per insect for pentatomids on vegetative and reproductive structures of their host plants, recorded using EPG

Stink bug species	N	Host plant	Feeding site	Recording time (hour)	Duration (min) feeding event
Edessa meditabunda	25 (25)[a]	Soybean stem	Xylem	8	61.6 ± 8.1
	25 (11)	Soybean stem	Phloem	8	99.8 ± 26.1
Nezara viridula	10 (10)	Soybean petiole	Xylem	9	25.6 ± 2.6
	10 (8)	Soybean pod	Xylem	9	55.0 ± 5.8
	10 (8)	Soybean pod	Seed endosperm	9	104.2 ± 25.1
Piezodorus guildinii	22 (21)	Soybean leaflet	Xylem	8	37.5 ± 4.7
	25 (25)	Soybean stem	Xylem	8	41.3 ± 4.7
	17 (17)	Soybean pod	Xylem	8	37.6 ± 5.3
	17 (6)	Soybean pod	Seed endosperm	8	80.2 ± 10.6
Diceraeus (*Dichelops*) *melacanthus*	21 (21)	Maize stem	Xylem	10	29.6 ± 3.7
Diceraeus (*Dichelops*) *furcatus*	10 (10)	Wheat stem	Xylem	8	24.9 ± 4.3
	10 (10)	Wheat seed head	Xylem	8	68.5 ± 13.9
	10 (10)	Wheat seed head	Seed endosperm	8	67.2 ± 9.6

Source: Lucini and Panizzi (2016, 2018), Mitchell et al. (2018)
[a]Values in parenthesis indicated the number of insects that exploited each feeding tissue
N number of insects recorded

egy, probably to obtain water, for a relatively short period of time (< 30 min) (Table 8.1). They also feed on stem parenchyma tissue using laceration and maceration tactics, and these cell rupture activities, although sometimes they are of short duration, are directly associated with the severe damage observed on seedlings.

In addition, when *D. furcatus* feed on wheat seed heads (seed endosperm), the feeding event is longer (ca. 70 min, which includes preparation plus ingestion of the cell contents) (Table 8.1). Again, the damage caused by the bugs on wheat plants, during reproductive stage, results from the long period of laceration/maceration activities. These damages include seeds malformed and discolored seed heads, symptom called as "white seed head" (Fig. 8.8), which is observed when bugs feed during the booting stage (Panizzi et al. 2016) (see Chap. 1, Sect. 1.6 and Chap. 4, Sects. 4.2.1 and 4.2.2 for more details).

These examples demonstrate that by relating waveforms generated with EPG to feeding sites revealed by histological sectioning, and considering the duration of the feeding events, one can explain the resulting damage to plants. In other words, EPG is a valuable tool to understand the relationship between true bug feeding and plant damage.

Fig. 8.8 Discolored (whitish) seed head of wheat cv. BRS Parrudo, caused by feeding of the stink bug *Diceraeus (Dichelops) furcatus* during booting stage, probably caused by extensive laceration/maceration activities. (Photo by AR Panizzi)

8.5 EPG and Integrated Pest Management (IPM)

Different strategies/tactics to manage true bug pest species and their interactions with the use of EPG were discussed previously (Chap. 7) and include insecticides (contact and systemic); parasitoids and endosymbionts; host plant resistance including traditional plant breeding and toxins expressed by transgenic plants; and gene silencing by RNAi technique.

In integrated pest management programs, EPG has been used in three main fronts: (1) understanding the causes and mechanisms involved in acquisition and inoculation of plant pathogens transmitted by hemipterans; (2) evaluating the effect of any chemical compound on feeding activities that are directly related to plant damage or transmission of a plant pathogen; and (3) assessing the effects of different plant varieties (including plants genetically modified to produce toxins) on the feeding behavior, in order to screen for resistant varieties.

Recently, Backus et al. (2021) discussed the fundamentals, controversies, and perspectives of EPG for arthropod pest management. These authors analyze other feeding guilds, such as blood-feeding arthropods and chewing insects and parasitoids; they also expand the use of EPG to explain mechanisms of crop damage, plant or animal pathogen transmission, and the effects of insecticides, antifeedants, repellents, or transgenic plants and animals, on specific behaviors related to damage or transmission.

Clearly, we are reaching a point of changing the more traditional and conservative uses of EPG (i.e., feeding of fragile piercing-sucking insects) to a new holistic scenario including not only other feeding guilds, but other behaviors as well. This change in approach will have many ramifications, which will include enhancing management strategies for pests within IPM programs.

8.6 Concluding Remarks

We have analyzed the importance of EPG as the most advanced electronic apparatus to monitor and reveal details of the feeding and other behaviors of true bugs. Furthermore, we have addressed how plant responses to feeding damage of true bugs can be better understood using the EPG as the "mediator" apparatus.

It is reasonable, however, to note that the use of EPG will not yet answer all the queries one may have relating to the feeding behavior of true bugs. We have discussed some limitations of the procedure and problems that may arise while conducting experiments on heteropterans. This, of course, is to be expected when using any equipment in a novel manner – remember that the use of EPG with true bugs falls, roughly, in a time frame of less than 30 years. Considering the latest advances in EPG technology, and the availability of monitors optimized for heteropteran research, that time frame reduces to less than 15 years. This is indeed a short time to fully develop any novel approach such as this one. As we analyze more species of true bugs on different host plants and obtain more results, the use of EPG will become more standardized. Therefore, the future of EPG in electronic monitoring of the activities of phytophagous true bugs is very promising, to enhance strategies/tactics to manage true bug pest species and to further our understanding of plant/bug interactions.

References

Backus EA, Guedes RNC, Reif KE (2021) AC-DC electropenetrography: fundamentals, controversies, and perspectives for arthropod pest management. Pest Manage Sci 77:1132–1149

Bernays E (2020) In praise of looking. Am Entomol 66:28–29

Bowling CC (1980) The stylet sheath as an indicator of feeding activity by the southern green stink bug on soybeans. J Econ Entomol 73:1–3

Cervantes FA, Backus EA, Godfrey L, Akbar W, Clark TL (2016) Characterization of an EPG waveform library for pre-reproductive adult *Lygus lineolaris* and *L. hesperus* feeding on cotton squares. Ann Entomol Soc Am 109:684–697

Depieri RA, Panizzi AR (2011) Duration of feeding and superficial and in-depth damage to soybean seed by selected species of stink bugs (Heteroptera: Pentatomidae). Neotrop Entomol 40:197–203

Hori K (2000) Possible causes of disease symptoms resulting from the feeding of phytophagous Heteroptera. In: Schaefer CW, Panizzi AR (eds) Heteroptera of economic importance. CRC Press, Boca Raton, pp 11–35

Krinski D (2013) First report of phytophagous stink bug in chicory crop. Cienc Rural 43:42–44

Krinski D, Pelissari TD (2012) Occurrence of the stinkbug *Edessa meditabunda* (F.) (Pentatomidae) in differents cultivars of lettuce *Lactuca sativa* L. (Asteraceae). Biosci J 28:654–659

Lucini T, Panizzi AR (2016) Waveform characterization of the soybean stem feeder *Edessa meditabunda* (F.) (Hemiptera: Heteroptera: Pentatomidae): overcoming the challenge of wiring pentatomids for EPG. Entomol Exp Appl 158:118–132

Lucini T, Panizzi AR (2017) Feeding behavior of the stink bug *Dichelops melacanthus* (Heteroptera: Pentatomidae) on maize seedlings: an EPG analysis at multiple input impedances and histology correlation. Ann Entomol Soc Am 110:160–171

Lucini T, Panizzi AR (2018) Electropenetrography (EPG): a breakthrough tool unveiling stink bug (Pentatomidae) feeding on plants. Neotrop Entomol 47:6–18

Lucini T, Panizzi AR (2020) Electropenetrographic (EPG) comparison of feeding behavior of *Dichelops furcatus* (F.) (Hemiptera: Heteroptera: Pentatomidae) on soybean and spring cereals. J Econ Entomol 113:1796–1803

Miles PW (1972) The saliva of Hemiptera. Adv Insect Physiol 9:183–255

Mitchell PL, Cooke SB, Smaniotto LF (2018) Probing behavior of *Nezara viridula* on soybean: characterization and comparison of electrical penetration graph (EPG) waveforms on vegetative and reproductive plant structures. J Agric Urban Entomol 34:19–43

Panizzi AR (1995) Feeding frequency, duration and preference of the southern green stink bug (Heteroptera: Pentatomidae) as affected by stage of development, age, and physiological condition. An Soc Entomol Brasil 24:437–444

Panizzi AR, Machado-Neto E (1992) Development of nymphs and feeding habits of nymphal and adult *Edessa meditabunda* (Heteroptera: Pentatomidae) on soybean and sunflower. Ann Entomol Soc Am 85:477–481

Panizzi AR, Niva CC, Hirose E (1995) Feeding preference by stink bugs (Heteroptera: Pentatomidae) for seeds within soybean pods. J Entomol Sci 30:333–341

Panizzi AR, Agostinetto A, Lucini T, Smaniotto LF, Pereira PRVS (2015) Management of green belly stink bugs *Dichelops* spp. on wheat. Embrapa Wheat, Series Doc 154:1–40 (*in Portuguese*)

Panizzi AR, Agostinetto A, Lucini T, Pereira PRVS (2016) Effect of green-belly stink bug, *Dichelops furcatus* (F.) on wheat yield and development. Crop Prot 79:20–25

Rizzo HFE (1971) Morphological and biological aspects of *Edessa meditabunda* (F.) (Hemiptera, Pentatomidae). Rev Per Entomol 14:272–281 (*in Spanish*)

Simmons AM, Yeargan KV (1988) Feeding frequency and feeding duration of the green stink bug (Hemiptera: Pentatomidae) on soybean. Environ Entomol 81:812–815

Todd JW, Turnipseed SG (1974) Effects of southern green stink bug damage on yield and quality of soybeans. J Econ Entomol 67:421–426

Wratten SD (1994) Video techniques in animal ecology and behaviour. Springer Science and Business Media, Dordrecht

Insect Index

A. R. Panizzi et al., *Electronic Monitoring of Feeding Behavior of Phytophagous True Bugs (Heteroptera)*, Entomology in Focus 6,
https://doi.org/10.1007/978-3-030-64674-5

Subject Index

A. R. Panizzi et al., *Electronic Monitoring of Feeding Behavior of Phytophagous True Bugs (Heteroptera)*, Entomology in Focus 6,
https://doi.org/10.1007/978-3-030-64674-5

X

Y

The manufacturer's authorised representative in the EU is Springer Nature Customer Service Centre GmbH, Europaplatz 3, 69115 Heidelberg, Germany. If you have any concerns regarding our products, please contact ProductSafety@springernature.com

Printed and bound by CPI Group (UK) Ltd, Croydon, CR0 4YY
30/09/2026
02225495-0003